Beyond the Visible: Predicting Metal Breakdown with Machine Learning

Sandeep

Table of Contents

Chapter 1: Introduction

This Ph.D. research has two main objectives. First, studying the nature of plasma membrane of eukaryotic cells, known as lipid bilayer, in aqueous environment using molecular dynamics (MD) simulations and machine learning (ML) methods. Second, investigating the application of ML for the purpose of modeling metallic corrosion. The rest of this document can be summarized as follows. In chapter 1, a brief introduction to MD and ML techniques is presented. Chapters 2 to 4 cover research objective I, while chapter 5 contains the details of research objective II. Finally, in chapter 6 the main conclusions of the research are reported.

Molecular Dynamics (MD) Simulations

MD is a technique to simulate molecular motion on a computer. Forces between atoms are electrostatic in nature because of interactions between electron clouds and nuclei. In classical MD, a representation of the physical system of interest is generated by placing N atoms/molecules in a volume V. The forces between the atoms are represented by mathematical functions of distance collectively called a force field. The nature of interactions between the atoms depends on their characteristics. In non-polar species, the dipole-dipole interactions between their electron clouds dominate, which are called van der Waals interactions. Van der Waals interactions are commonly represented by Lennard-Jones force field. Charged and polarized species interact via Coulombic interactions. Bond and angular interactions are represented by harmonic potentials. In addition, potential functions are also defined for dihedral interactions within a molecule. **Equation 1** shows the force field or potential that defines the energies between particles.[1]

$$V(r) = \sum_{bonds} K_r\left(r - r_{eq}\right)^2 + \sum_{angles} K_\theta\left(\theta - \theta_{eq}\right)^2$$
$$+ \sum_{dihedral} \frac{V_n}{2}[1 + cos(n\varphi - \gamma)]$$
$$+ \sum_{i<j} \left[\frac{A_{ij}}{R_{ij}^{12}} - \frac{B_{ij}}{R_{ij}^6}\right] + \sum_{i<j} \frac{q_i q_j}{\in R_{ij}} \qquad \text{Equation 1}$$

The first term represents a harmonic bond between two bonded particles with a spring constant K_r and equilibrium distance of r_{eq}. The second term is the harmonic angle between three bonded particles with a constant K_θ and equilibrium angle of θ_{eq}. The third term is the dihedral angle that are between four particles and represent a potential of rotation about the central bond. The fourth and fifth terms are space energies of Van der Waals and electrostatics between non-bonded particles. The negative gradient of the potential energy is the force. Once the forces acting on all the particles are calculated, the motion of the particles is generated by numerically integrating the classical Newton's equation of motion. By doing this for long enough time until reaching to the equilibrium we can analyze behavior of different systems under different conditions. Often, it is useful to represent molecules using a coarse-grained description wherein all a group of atoms are represented by a united bead. A popular coarse-grain model to study lipids is called the MARTINI force field.[2]

Machine Learning (ML) Algorithms

Machine learning methods are suitable for developing predictive models in the cases where a large dataset is available, the outcome to be predicted depends on several variables, and when a mechanistic model of the relationship between the input variables and the outcome is not well established. Before we embark on describing our work in detail, it is useful to provide a brief introduction to machine learning methods used here for the readers who are experts in the corrosion field but not necessarily in ML. Machine learning refers to a class of algorithms that have the ability to learn and perform a task when they are provided with some data.[3,4] Thus, ML algorithms get to learn how to perform a task, such as predicting the outcome of an experiment, when they are trained on the data obtained from some previously performed experiments. Specifically, the ML algorithms that learns from data wherein the outcome of the experiment is specified are called supervised learning.[4] Some popular supervised ML algorithms are Random Forest (RF), Artificial Neural Network (ANN), Supported Vector machines Regression (SVR), and K Nearest Neighbors (KNN).

To understand RF algorithm, one needs to first understand what is meant by a decision tree. In a decision tree, a dataset is split around input variables into smaller subsets. The split is performed so that the subsets that are formed have smaller variances in the outcome values. Each split can be thought of as a branch of the tree and each data subset as a leaf. The data are progressively split until some terminal condition is met. The terminal condition can be either that the maximum number of splits has been performed or the standard deviation of a subset has fallen below a cutoff value. The average value of the outcome in the terminal leaf, that is the leaf, which is not split any further, is the predicted value of the outcome for those set of input variables. As a simple example, imagine a decision tree which is used to predict the corrosion rate as a function of two key variables, pH and temperature, is created with a tree split with branches pH < 7 and pH $\geq$ 7. These branches are each further split for temperature < 350 K and temperature $\geq$

350 K. Then, to predict the corrosion rate for a condition, say with pH = 4.5 and temperature = 300 K, the average value of the corrosion rate for the leaf: pH < 7 → temperature < 350 K is the predicted value of this decision tree. A RF is comprised of a multitude of decision trees. The decision trees are formed based on random subsets of the training data. RF algorithm reports the mean value of the predictions from all decision trees. In general, using the outcome of many ML models to make the final prediction is called ensemble learning, and has been shown to significantly improve the prediction performance.[5] Therefore, RF is an ensemble learning method based on numerous decision trees. So, in the example, the corrosion rate is the output variable (label) and pH and temperature are the input variables (features).[6]

ANNs are networks of interconnected nodes that act as universal approximators, that is, they can approximate any continuous function to an arbitrary level of accuracy with a finite number of nodes.[7] The architecture of ANN is as follows: In the first layer (named input layer), each input variable, x_i is fed to a node. Each node in the input layer is connected to the nodes in the next layer (called first hidden layer). The connections between the nodes are assigned some weights, w_{ij}. At each node in the first hidden layer, a weighted sum of the inputs from the input layer is calculated, $F_j = \sum_i w_{ij} x_i$. The F_j is transformed via an activation function, such as a sigmoidal function, which is the output from each node, $x_j = \frac{1}{1+e^{-F_j}}$, which becomes the input from node j in the first hidden layer to all the nodes in the second hidden layer. This process is repeated for all the layers until the output layer is reached, which in regression problems is a single node that predicts the outcome. Training an ANN architecture entails adjusting the weights connecting the nodes so as to minimize the mean squared error between the predicted and the actual outcome values/classes.[8,9]

SVR is another powerful supervised learning algorithm. In general, the relationship between the input variables and the label is non-linear. In SVR, the input variables are transformed to higher dimensions where the relationship is linear. In a higher dimension, a linear regression line is fitted to the data. For example, a polynomial relationship, such as $y = a_0 + a_{11}x_1 + a_{12}x_2 + .. + a_{n1}x_1^n + a_{n2}x_2^n$ can be represented as a *2n* dimensional hyperplane. There are special functions called kernels that help in determining the hyperplane in the higher dimension without increasing the computational cost.[10]

KNN is a simple algorithm in which, for each data point, K nearest-neighbors are identified in the training set (where K is an integer), and the average value of their labels is reported as the outcome of that data point.[11] The K nearest-neighbors are identified by defining a *distance*. For numerical input variables, Euclidean distance is commonly used, whereas, for categorical variables, Hamming distance is used.[12]

As discussed above, every ML algorithm has a specific architecture. For instance, an ANN is characterized by the number of hidden layers and the number of nodes per hidden layer; a RF is characterized by the number of trees and the maximum number of features that can be split. These parameters that define the architecture of a ML algorithm are called hyperparameters.[13] The performance of a ML algorithm on a dataset varies as one changes the hyperparameters. Therefore, values of the hyperparameters need to be tuned/adjusted to optimize performance. To tune the hyperparameters, first, the data should be split into a training set and a testing set. The training set is often comprised of 70-80% of the entire dataset. The optimum values of the hyperparameters are found by evaluating the performance of all formed ML models on the training set. Once the hyperparameters are fixed, the founded best ML algorithm is trained on the training set, and then its performance is evaluated on the testing set.

Chapter 2: Relationship Between Cholesterol Distribution and Ordering of Lipids

The material presented in this chapter is put together as a paper: "Aghaaminiha, M.; Farnoud, A. M.; Sharma, S. Quantitative Relationship between Cholesterol Distribution and Ordering of Lipids in Asymmetric Lipid Bilayers. *Soft Matter* **2021**, *17* (10), 2742–2752".[14]

Introduction

The plasma membrane is a lipid bilayer that separates the cell from the outside environment.[15–17] While different mammalian cells are known to have a wide variety of lipids in their plasma membrane, it is now well-understood that the phospholipid composition of the inner (cytofacial) leaflet in all mammalian cells is different from that of the outer (exofacial) leaflet.[18–20] Verkleij et al. pioneered studies on phospholipid asymmetry, showing that sphingomyelin (SM) and phosphatidylcholine (PC) are abundant in the outer leaflet, while phosphatidylethanolamine (PE) and phosphatidylserine (PS) are primarily localized in the inner leaflet.[21] These findings were later verified by the recent studies of Lorent et al.[19] and Vahedi et al.[22] The phospholipid asymmetry in membrane leaflets has important functional implications for the cell. For instance, in healthy red blood cells, PS is known to be almost exclusively localized in the inner leaflet, the presence of this lipid in the outer leaflet is a sign of eryptosis and a signal to macrophages to phagocytose the old red blood cells and maintain homeostasis.[18,23]

While the asymmetry of phospholipids in the plasma membrane is well-established, there is little consensus regarding cholesterol distribution in different leaflets of the plasma membrane.[18,20,24] Unlike phospholipids, there are no specific enzymes for cholesterol transport across the bilayer, and cholesterol distribution is believed to be primarily dictated by its affinity towards other lipids.[18,24] However, experimental results in this regard have been contradictory. It has been reported that cholesterol molecules are enriched in the outer leaflet because of their strong interactions with the saturated acyl chain phospholipids, specifically SM.[25–27] Contrarily, it has also been suggested that the inner leaflet is favored by cholesterol due to (i) the abundance of PE phospholipids in this leaflet which provides higher negative curvature,[28,29] and (ii) the abundance of C24 sphingolipids in the outer leaflet which pushes cholesterol to the inner leaflet.[30] There are also other reports that mention that cholesterol is symmetrically distributed in the cell membrane.[31,32]

Experimental studies of cholesterol distribution are complicated by several confounding factors. First, different mammalian cells have different membrane lipid compositions.[33,34] Secondly, experimental methods such as quenching of dehydroergosterol fluorescence, filipin staining, noninvasive neutron scattering, and enzymatic degradation by cholesterol oxidase, while helpful, each suffer from various drawbacks.[35,36] Finally, cholesterol is known to flip-flop between the two leaflets at a rapid rate, [37–40] making it difficult for experimental probes to trace it with high accuracy at any time-point.[41] A mechanistic study on how changes in lipid structure and chemistry affect cholesterol distribution helps to provide a general understanding of cholesterol asymmetry in the plasma membrane.

In the current work, we have studied the spatial distribution of cholesterol within asymmetric lipid bilayers via coarse-grained (CG) molecular dynamics (MD) simulations. We have performed two sets of simulations, with the first focusing on characterizing cholesterol distribution in a lipid bilayer mimicking the known membrane lipid composition of red blood cells and the second focusing on elucidating how the structure and chemistry of the lipids in the bilayer affect cholesterol distribution. Results reveal that the ordering of lipids in membrane leaflets is the primary regulator of cholesterol spatial distribution in the asymmetric bilayers.

Methodology

Force Field

We have employed the coarse-grained (CG) Wet Martini force field with polarizable water to model all lipids, cholesterol, ions, and water molecules in the system. The polarizable water model for the Martini force field, developed by Yesylevskyy et al.,[42,43] is a three-bead model to represent four water molecules and can capture the orientational polarizability and the dielectric screening effect of bulk water. The topologies of lipids are described in **Figures A1-C** and **A2-B** (Appendix A) alongside the name and type of each bead. The main types of particles in the Martini CG force field are P, polar; N, intermediate; C, apolar; and Q, charged. Subscripts *0*, *a*, *d*, and *da* indicate if the particle has hydrogen-bond (H-bond) forming capacities. The particle with "*0*" has no H-bond capacity, "*a*" is a H-bond acceptor, "*d*" is a H-bond donor, and "*da*" is both H-bond acceptor and donor.[44] Other subscripts are numbers: 1, 2, 3, 4, and 5 that display the degree of polarity, where 5 is the most and 1 is the least polar particle.[44] Periodic boundary conditions were applied in all three directions of the simulation box. In the first set of simulations (with the sample snapshot in **Figure A1-B**; Appendix A), the simulation box size was $250 \times 250 \times 125$ (Å^3), the total number of lipids in the systems was 2,043, and the total number of water molecules was 41,487. In the second set of simulations (with the sample snapshot presented in **Figure A2-A**; Appendix A), the simulation box size was 125 Å in all directions, the total number of lipids in the systems was 510, and the total number of water molecules was 10,834. Initial configuration of lipid bilayers was generated using the *Insane.py* script developed by Marrink et al.[45] All simulations were performed on the GROMACS/5.1.2 molecular simulation package.[46]

Simulation Details

The initial configuration of the system was energy minimized by employing steepest descent algorithm with the maximum force tolerance of 1000.0 $kJ.mol^{-1}.nm^{-1}$. This was followed by a 25 ns canonical ensemble (constant number of particles N, volume V, and temperature T) simulation and a 25 ns isothermal-isobaric (constant N, pressure P, T) ensemble simulation for pre-equilibration. The time-step chosen for both the ensembles was 10 fs. Then, three replicas of 10 µs NPT simulation with a timestep of 20 fs were performed, as in previous studies [17], and the data was collected for the last 6 µs for all the simulations.

The parameters of the simulation system parameters were chosen to be the same as in the previous work done to parameterize the polarizable water model on the Martini force field.[43] Briefly, a spherical cut-off of 1.1 nm was chosen for the Lennard Jones (LJ) interactions. For Coulombic interactions, Reaction-Field-zero (RF-zero) was employed with the spherical cut-off of 1.1 nm.[43] The RF-zero and PME (Particle Mesh Ewald) electrostatics have been systematically compared previously and no differences in the bilayer properties were identified.[43] The relative dielectric constant was taken to be

2.5.[42,43] Both LJ and Coulombic interactions were smoothly shifted to be zero beyond the

cut-off. In the short pre-equilibration runs, Berendsen thermostat and barostat with the

time constants of 1 ps and 3 ps, respectively were employed for quick equilibration. For

the NPT equilibration and production runs, the velocity-rescaling thermostat developed

by Bussi et al.[47] was used. Semi-isotropic pressure coupling was accomplished using

Parrinello-Rahman barostat with the coupling time constant of 12 ps.[17,43] Along the axis

of the bilayer plane and the z-direction, the reference pressure and isothermal

compressibility were set to be 1.0 bar and 3×10^{-4} bar^{-1}, respectively. All simulations are

performed at 37°C.

Radial distribution functions, g(r), were calculated to examine the proximity of

the hydroxyl headgroup of cholesterol (ROH CG bead) to various chemical moieties in

phospholipids. Pioneering work by Huang has described the importance of the proximity

of the –OH group of sterols, in the β configuration, to the carbonyl oxygen of the fatty

acyl groups in phospholipids for strong cholesterol-lipid interactions.[48]

We divide the bilayer into three regions, namely, inner leaflet, midplane, and

outer leaflet. The threshold for separating these regions was set to be 0.8 nm from the

center of the bilayer as previously suggested.[16,17] Cholesterol flip-flop rate was defined as

the average frequency by which the cholesterol molecules undergo intramembrane

exchange, in other words, leave the inner/outer leaflet, reach the outer/inner leaflet, and

then return to the inner/outer leaflet.[37,38,49] Orientation of a cholesterol molecule within a

bilayer was defined as the angle between vector, $\bar{a}$ (the vector joining the ROH bead to the C1 bead of cholesterol, see **Figure A1-C,** Appendix A) and vector $\bar{z}$ (the bilayer

normal). To measure the ordering of lipids in each leaflet, ensemble-averaged order

parameter of the acyl chains of lipids[2,50] was calculated , as in **Equation 2**:

$$S_{Chain} = \langle \frac{1}{2}(3\langle cos^2\theta\rangle - 1)\rangle \qquad \text{Equation 2}$$

Where θ is the angle between each bond vector of the acyl chain beads and the bilayer

normal. A high value of the S_{Chain} (~ 1) indicates that the bonds are aligned with the

bilayer normal, whereas a low value (~ 0) indicates that the bonds are randomly oriented

respect to the bilayer normal.[2,50] The area per lipid, *a(x)* for each leaflet was calculated

based on the ensemble-averaged area of the plane in the simulation box, *A(x)* divided by

the total number of two-chain lipids + cholesterol molecules in that leaflet according to

Equation 3 (*x* is the mole fraction of cholesterol in the system).[51]

$$a(x) = \frac{A(x)}{N_{lipids} + N_{chol}} \qquad \text{Equation 3}$$

The thickness of the bilayer was estimated as the ensemble-averaged distance

between the phosphate head group of the lipids (PO4 CG bead shown in **Figure A1-C,**

Appendix A) in the two leaflets.[51,52]

Results and Discussion

Cholesterol Distribution in the Leaflets of the Bilayer

The asymmetric distribution of lipids in the plasma membrane is expected to

affect the localization of cholesterol in membrane leaflets. To examine this phenomenon,

an asymmetric lipid bilayer, mimicking the lipid composition of the plasma membrane of

red blood cells was studied. The asymmetric lipid bilayer comprised of seven different lipids. The outer leaflet lipids were selected to be POPC/16:0SM/POPE/CHOL with the molar ratio of 31/29/7/33 while the inner leaflet lipids were DOPC/16:1SM/DOPE/DOPS/CHOL with the molar ratio of 10/7/31/19/33. This composition of phospholipids matches the predominant lipids in each leaflet of the plasma membrane of red blood cells.[19,21,22] Cholesterol was the only sterol included in the system with a molar ratio of 1:2 with respect to the total number of lipids in the two leaflets. Cholesterol content has been reported to be in the range of 30 mol% to 50 mol% in mammalian cell plasma membranes.[53,54] Two symmetric model lipid bilayers were also studied. In the first bilayer, labeled cyto-symmetric, the two leaflets had the same lipid composition as that of the inner (cytofacial) leaflet of the asymmetric bilayer noted above. In the second symmetric bilayer, labeled exo-symmetric, the composition of both leaflets was the same as the outer (exofacial) leaflet of the asymmetric bilayer. **Table 1** listed the lipid compositions of the three systems.

Table 1

Composition of the inner and outer leaflets of the studied lipid bilayers
Cholesterol was included in all the bilayers with the molar ratio of 1:2 with respect to the
total number of lipids.

System	Inner leaflet	Outer leaflet
Asymmetric	DOPC/16:1SM/DOPE/DOPS/CHOL: 10/7/31/19/33	POPC/16:0SM/POPE/CHOL: 31/29/7/33
Cyto-symmetric	DOPC/16:1SM/DOPE/DOPS/CHOL: 10/7/31/19/33	DOPC/16:1SM/DOPE/DOPS/CHOL: 10/7/31/19/33
Exo-symmetric	POPC/16:0SM/POPE/CHOL: 31/29/7/33	POPC/16:0SM/POPE/CHOL: 31/29/7/33

In the initial configuration, cholesterol molecules were equally distributed between the two leaflets. Upon equilibration of the simulation, cholesterol distribution became asymmetric in the bilayer with asymmetric lipid composition, and symmetric in case of the two symmetric bilayers, confirming the direct role of phospholipid chemistry in cholesterol localization (**Figure 1A**). In the asymmetric bilayer, the fraction of cholesterol molecules in the inner and outer leaflets were 41±1% and 53±2%, respectively, while 6±2% of cholesterol resided in the midplane of the bilayer. In the cyto-symmetric bilayer, mimicking the inner leaflet, a larger fraction of cholesterol was found in the midplane (11±3% compared to 6±2% for the asymmetric bilayer), while the exo-symmetric bilayer only had 3±1% of cholesterol molecules in the midplane. This result is consistent with the experiments showing that cholesterol have a preference for the bilayer midplane in the case of polyunsaturated bilayers.[55]

Figure 1B shows the orientation of cholesterol molecules as a function of location within the bilayer for the three bilayers. For this calculation, the lipid bilayer was divided into ~30 slabs parallel to the plane of the bilayer. In each slab, the ensemble-averaged angle of the cholesterol molecules with respect to the bilayer normal was found. The reported angle with respect to the bilayer normal corresponds to the average over the configurations in the last 6 μs of three independent replicas of each system. Within 0.8 nm from the center of the bilayer, cholesterol orientation changes almost linearly with the distance from the center for all the three studied systems. At the distance of 0.8 nm from the center, a sharp change in the orientation of cholesterol is observed and so this region is defined as the midplane.[16]

Figure 1

Percentage and orientation of cholesterol in different regions of the bilayers
A) Percentage of cholesterol molecules located in the inner leaflet (cyan), midplane (black striped), and the outer leaflet (purple) of the asymmetric, cyto-symmetric, and exo-symmetric lipid bilayers. B) Ensemble-averaged angle of cholesterol molecule with the bilayer normal as a function of the location of the hydroxyl headgroup of cholesterol. Shaded area represents the midplane of the bilayer. Beyond the midplane, the orientation is almost independent of the location. Error bars in both the figures were estimated as standard error of the mean when the last 6 μs of the three replicas of each system were split into four equally sized blocks of 1.5 μs each.

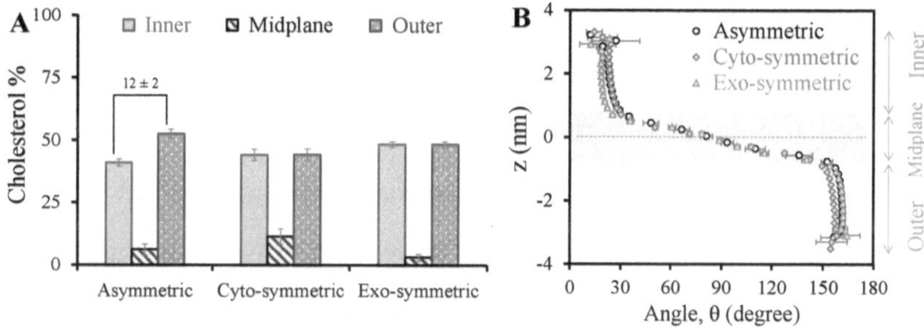

It has been suggested that the alignment of cholesterol in the leaflet favors the cholesterol-lipid interaction in the PM.[48,56] Therefore, it is hypothesized that the ordering of lipids in a leaflet should favor the presence of cholesterol. Indeed, the outer leaflet of the asymmetric bilayer has a larger S_{Chain} (0.48±0.02) as compared to that of the inner leaflet (0.33±0.02). S_{Chain}, as defined in **Equation 2**, is a measure of ordering of the lipids in a leaflet; a high value of S_{Chain} corresponds to more order. The exo-symmetric and the cyto-symmetric bilayers have the S_{Chain} values of 0.48±0.02 and 0.32±0.02, respectively, for both leaflets, which explains the higher concentration of cholesterol in the exo-symmetric leaflets as compared to in the cyto-symmetric leaflets. These findings suggest that the ordering of the lipids in the leaflets is an important factor in determining the localization of cholesterol within the leaflets of the bilayer. Other physical characteristics of these bilayers, such as area per lipid and the thickness, were also compared and are listed in **Table A1** (Appendix A). The cyto-symmetric is the thinnest and exo-symmetric is the thickest bilayer, as expected, but the difference in the thickness is only ~ 2 Å. The asymmetric bilayer has a slightly smaller area per lipid (0.52 ± 0.01 nm^2) at the inner leaflet as compared to the cyto-symmetric bilayer (0.52 ± 0.01 nm^2) and has slightly larger area per lipid (0.48 ± 0.01 nm^2) at the outer leaflet as compared to the exo-symmetric bilayer (0.47 ± 0.01 nm^2). This difference confirms that the asymmetry influences the properties of the leaflets.[57] The values of area per lipid are smaller than those usually reported for the bilayers where no cholesterol is present (~ 0.5 to 0.7 nm^2).[58]

It should be noted that the composition of lipids in different leaflets of the asymmetric bilayer did not get altered during the simulations. The density profiles of different lipids in the equilibrated bilayer confirm this point (**Figure A3**; Appendix A). On average, in the asymmetric bilayer only 2 out of 103 DOPC, 1 out of 74 SM, 7 out of 311 DOPE, and 6 out of 192 DOPS lipids moved from the inner leaflet to the outer leaflet, while no lipid molecule from the outer leaflet moved to the inner leaflet.

Cholesterol Flip-Flop Rate

A large range of cholesterol flip-flop rates have been reported in previous simulations[37,38] and experiments.[36,59,60] It has been argued that chemical probes or markers that are employed for experimental measurements of the flip-flop rates influence the rates.[36] While some non-invasive but indirect measurements of the flip-flop rates[36] have been introduced, the predictions made from these methods have remained contentious.[35,61] Cholesterol flip-flop rate is understood to increase with temperature and the amount of unsaturated lipids[62] and decrease with the thickness of the bilayer.[23] However, there is little understanding of how the flip-flop rates are affected by changes in bilayer (a)symmetry. One would presume that the flip-flop rate in an asymmetric bilayer will be dictated by the rate determining step that is the slower flip-flop rate observed for the more ordered leaflet. Furthermore, previous investigations have not studied the distribution of cholesterol flip-flop rates. **Figure 2** shows the distribution of cholesterol flip-flop rates for the asymmetric, cyto-symmetric, and exo-symmetric bilayers. The more ordered exo-symmetric bilayer showed the lowest average flip-flop rate of 0.77 ± 0.03 μs^{-1}. The flip-flop rate of the asymmetric bilayer was found to be higher, 1.39 ± 0.04 μs^{-1}. The flip-flop rate of the cyto-symmetric bilayer was the highest at 2.80 ± 0.05 μs^{-1}. Interestingly, the distribution of flip-flop rates was symmetric around the mode value for cyto-symmetric bilayer, whereas for both the asymmetric and the exo-symmetric bilayer, the distribution was skewed. An interesting conclusion is drawn from this analysis that the cholesterol flip flop rates of asymmetric lipid bilayers is in-between the flip flop rates observed for symmetric bilayers made up of their individual leaflets.

Figure 2

Distribution of cholesterol flip-flop rates for the three bilayers
Exo-symmetric bilayer with more ordered lipids has a narrower distribution with the smallest mode value. Cyto-symmetric bilayer has a broader distribution with the largest mode value. The flip-flop rate of the asymmetric bilayer is in-between the two symmetric bilayers. The error bars are estimated as the standard error of the mean when the last 6 μs of the simulations of three replicas of each system were split into four equally sized blocks.

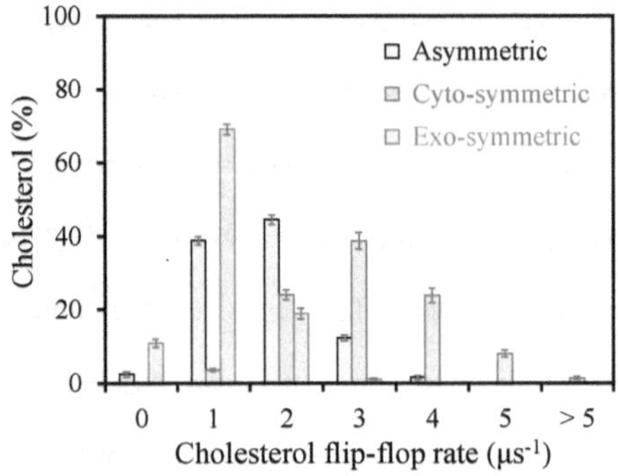

Arrangement of Cholesterol in the Plane of the Bilayer

Next, we analyzed the spatial arrangement of cholesterol and phospholipids within the leaflets of the asymmetric lipid bilayer via 2D radial distribution functions, $g_{2D}(r)$. **Figure 3 (A** and **B),** show that the ROH-AM and ROH-GL $g_{2D}(r)$ display a strong peak at the location of first coordination shell. First coordination shell can be thought of as the closest distance around a molecule that is accessible to other molecules. Therefore, this result suggests that there is a strong affinity of cholesterol molecules for the amide and glycerol beads of phospholipids (refer to **Figure A1-C**, Appendix A, for the representation of different CG Martini beads of the lipids and cholesterol).

The 2D RDF between cholesterol molecules, the ROH-ROH $g_{2D}(r)$, shows that cholesterol molecules do not cluster together but rather are separated by phospholipids molecules. This observation is in accordance with the "umbrella model" that postulates that cholesterol molecules are hidden from water by the phospholipid head groups. If cholesterol molecules were to cluster together then they would get exposed to water.[63,64] **Figure 3C** shows the three-dimensional radial distribution function, g(r) between the water molecules and the headgroup of the phospholipids and cholesterol. This g(r) shows that the headgroups of the phospholipids are solvated by water, whereas those of cholesterol are hidden from water, further confirming the umbrella model.

Figure 3

*2D and 3D radial distribution functions of different beads of lipids around the hydroxyl headgroup of cholesterol (ROH CG bead) at the **A)** inner leaflet and **B)** outer leaflet of the asymmetric bilayer. ROH headgroup of cholesterol resides near the amide (AM) and glycerol (GL) beads of lipids. On the other hand, no significant local structure is observed between ROH and other beads (PO4, phosphate; NC3, choline; NH3, ethanolamine; and CN0, serine). **C)** 3D radial distribution function, g(r) of water around the headgroup beads of lipids for the asymmetric system. CN0, NC3, and NH3 represent the headgroup of PS (phosphatidylserine), PC (phosphatidylcholine) & SM (sphingomyelin), and PE (phosphatidylethanolamine) phospholipids, respectively, and ROH represents the hydroxyl headgroup of cholesterol.*

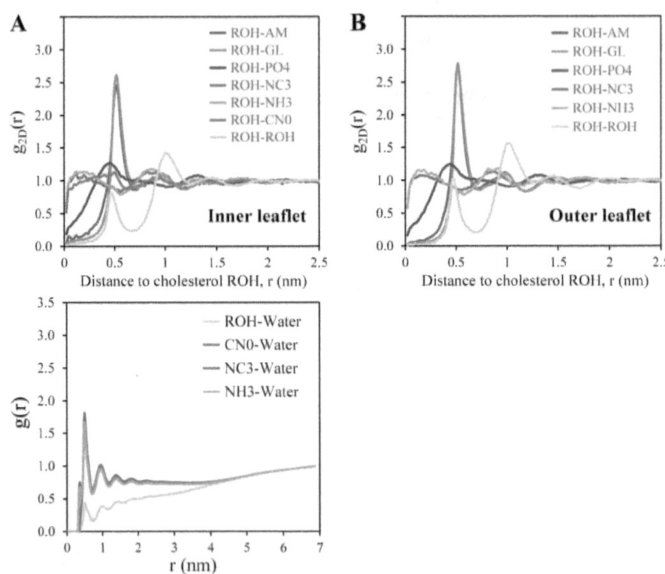

Cholesterol Localization as a Function of lipids Structural Properties

Our simulations of asymmetric lipid bilayer mimicking those of red blood cells showed that cholesterol distributes unequally between the two leaflets. To investigate the underlying reasons that govern cholesterol distributions in lipid bilayers, we studied various asymmetric bilayers comprising of two kinds of lipids. In these bilayers, one leaflet was always kept as DPPC, while the lipids of the other leaflet were varied. For the sake of convenience of terminology, the DPPC leaflet was labeled as the "outer" leaflet and the other leaflet as the "inner" leaflet. The types of lipids in the inner leaflet were selected so as to isolate the effect of lipids' backbone, headgroup, acyl chain saturation, and acyl chain length on cholesterol distribution. The compositional details of these lipid bilayers are listed in **Table 2**. The snapshot of DPPEi/DPPCo/CHOL lipid bilayer is displayed in **Figure A2** (Appendix A).

Table 2

Composition of asymmetric bilayers with DPPC in the outer leaflet
Cholesterol was present in all the bilayers with a molar ratio of 1:2 with respect to the
total number of lipids. The "i" and "o" subscripts represent the inner and outer leaflets.

Feature	Composition	Headgroup	Backbone	Saturation	Chain length LIPIDi/LIPIDo
Control	DPPCi/DPPCo/CHOL	choline	glycerol	saturated	16-16/16-16
Backbone	16:0SMi/DPPCo/CHOL	choline	sphingosine	saturated	16-16/16-16
Headgroup type	DPPEi/DPPCo/CHOL	ethanolamine	glycerol	saturated	16-16/16-16
	DPPSi/DPPCo/CHOL	serine	glycerol	saturated	16-16/16-16
Acyl chain saturation	DOPCi/DPPCo/CHOL	choline	glycerol	unsaturated	18-18/16-16
	POPCi/DPPCo/CHOL	choline	glycerol	mono-unsat.	16-18/16-16
Acyl chain length	12:0PCi/DPPCo/CHOL	choline	glycerol	saturated	12-12/16-16
	24:0PCi/DPPCo/CHOL	choline	glycerol	saturated	24-24/16-16

As before, all the simulations were performed at 37°C; the molar ratio of cholesterol to the total number of lipids was kept 1:2; and cholesterol molecules were equally divided between the two leaflets initially. As a control, a symmetric

DPPCi/DPPCo/CHOL system was also examined. For this system, as expected, cholesterol was equally distributed between the two leaflets (**Figure 4A**, left panel).

The effect of the lipid backbone. The role of lipid backbone in modulating cholesterol distribution was studied from a system containing cholesterol, 16:0SM in the inner leaflet, and DPPC in the outer leaflet (16:0SMi/DPPCo/CHOL). The headgroup of SM and DPPC is the same, but their backbone is different. SM has polar moieties in the backbone (AM1 and AM2), which are missing in DPPC. It was observed that cholesterol molecules have a higher concentration in the 16:0SM leaflet compared to the DPPC leaflet (**Figure 4A**, right panel). It is understood that the sphingosine backbone containing the amide groups have a stronger interaction with the headgroup of cholesterol, which agrees well with the experimental studies.[25,26]

The effect of the lipid headgroup. The effect of the lipid headgroup was studied in a lipid bilayer with the outer leaflet as DPPC and the inner leaflet as DPPE or DPPS. It was observed that cholesterol has a stronger affinity for the lipids with serine (DPPS) and ethanolamine (DPPE) headgroups than the lipids with a choline headgroup (DPPC) (**Figure 4B**). This finding is in agreement with the suggestion that PE, which is known to predominantly exist in the inner leaflet of the PM,[19,21,22] draws cholesterol to that leaflet (for complete discussion see[29]).

The effect of the acyl chain saturation. Next, the distribution of cholesterol was examined in two bilayers: with DPPC in the outer leaflet, the first bilayer had DOPC (with two unsaturated acyl chains), and the second bilayer had POPC (with one unsaturated acyl chain) in the inner leaflet. In both cases, cholesterol was found to localize in the leaflet with more saturated lipids (**Figure 4C**), in agreement with previous reports.[56,62,65,66] Importantly, the difference in cholesterol distribution between the two leaflets of these systems was much more significant than the other systems studied by us. In accordance with previous studies,[55] we find higher amount of cholesterol in the mid-plane for the DOPCi/DPPCo bilayer as compared to the POPCi/DPPCo bilayer as DOPC has higher degree of unsaturation than POPC. One explanation is that unsaturated lipids with disordered acyl chains do not allow proper alignment of cholesterol in the leaflet.

The effect of the acyl chain length. To examine the role of lipid chain length in cholesterol distribution, bilayers containing DPPC in the outer leaflet and 12:0PC or 24:0PC in the inner leaflet were studied (**Figure 4D**). In both cases, cholesterol was preferentially localized in the leaflet that had lipids with longer acyl chain length. A previous study by Courtney et al. reported that cholesterol is enriched in the inner leaflet where shorter lipids are located as compared to the outer leaflet where long chain sphingolipids exist.[30] While in our study, the inner and the outer leaflets have the same headgroup, backbone, and chain saturation, Courtney et al. used a variety of different lipids in the two leaflets, and thus did not isolate the effect of the acyl chain length.

Figure 4

Cholesterol distribution within different asymmetric bilayers
(Inner leaflet: cyan, midplane: black striped, and the outer leaflet: purple). The effect of
lipids' A) backbone, B) headgroup, C) acyl chain saturation and D) acyl chain length on
cholesterol distribution is analyzed. In all the systems, the outer leaflet was taken as
DPPC, and the lipid type of the inner leaflet was varied.

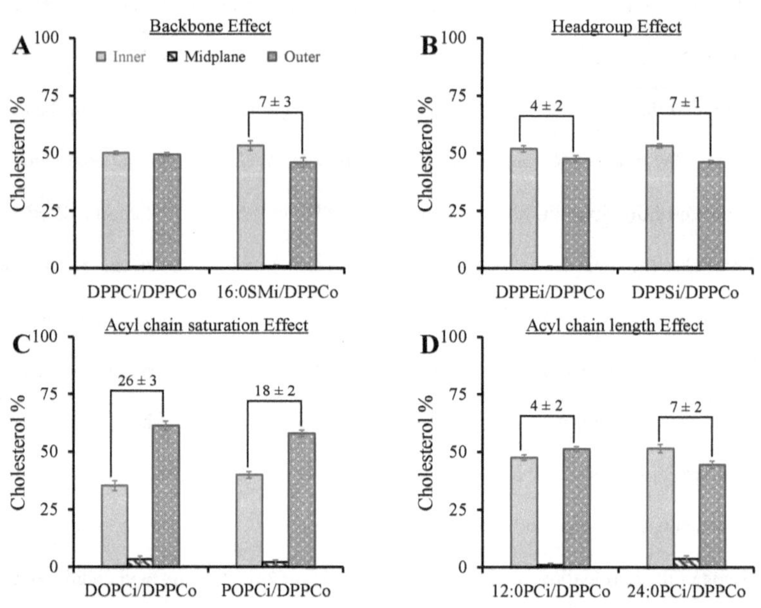

From these studies, it was found that acyl chain saturation results in the strongest
asymmetrical distribution followed by the lipid backbone, the headgroup type, and the
acyl chain length. Characteristics of these bilayer systems, such as the area per lipid,
bilayer thickness, S_{Chain} and cholesterol flip-flop rates are tabulated in **Table A2**
(Appendix A). To understand the relationship between cholesterol distribution and the

ordering of lipids in the bilayer, the ratio of the distribution of cholesterol in the inner and the outer leaflets versus the ratio of the S_{Chain} of the lipids in the inner and the outer leaflets was plotted (**Figure 5A**). Interestingly, for all bilayers, the data lies close to the y = x line, implying that the ratio of distribution of cholesterol is similar to the ratio of the ordering of lipids in the leaflets. This striking result suggests that there is a unifying principle governing the distribution of cholesterol in lipid bilayers. It is well understood that cholesterol promotes ordering of lipids in the leaflets. Therefore, it is like a chicken-or-egg conundrum as to whether the ordering of the lipid bilayers governs the observed cholesterol distribution or is it the distribution of cholesterol that results in the ordering of the lipids. To answer this question, we performed simulations of the asymmetric lipid bilayers without any cholesterol to calculate their S_{Chain}. In **Figure 5B** we have plotted the ratio of cholesterol distribution in these bilayers obtained previously with respect to the ratio of the S_{Chain} of the leaflets when cholesterol was absent. A positively sloped linear relationship between the distribution of cholesterol and the ordering in the two leaflets (in the absence of cholesterol) is observed. This result suggests that the distribution of cholesterol is dictated by the relative ordering of the lipids. Certainly, the presence of cholesterol increases the S_{Chain}, but it is not the sole effect underlying the relationship observed in **Figure 5A**. The S_{Chain} of the lipids in both leaflets in the presence and absence of cholesterol is summarized in **Table A3** (Appendix A).

Figure 5

Fraction of cholesterol in the leaflets vs. average order parameter of lipids
A) Relationship between cholesterol localization in the leaflets with respect to the average order parameter (S_{Chain}) of lipids. The ratio of cholesterol in the inner and outer leaflets lies close to the ratio of S_{Chain} of lipids in the inner and outer leaflets. B) Ratio of fraction of cholesterol in the inner and outer leaflets for different lipid bilayers plotted as a function of the ratio of the S_{Chain} of lipids in the same bilayers but in the absence of cholesterol. A positively sloped linear relationship is observed. This figure corresponds to the systems with the same number of two-chain lipids in the inner and outer leaflets.

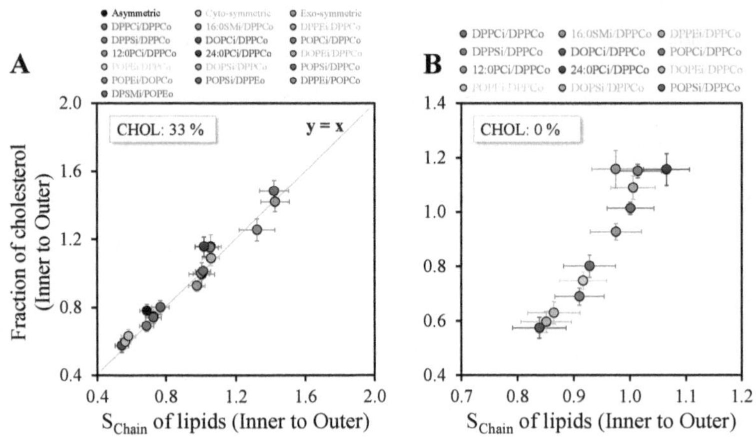

To explain the distribution of cholesterol observed in the equimolar asymmetric bilayers, we studied symmetric bilayers comprised of the lipid types studied above with a mole fraction of cholesterol equal to 0.33. **Table A4** (Appendix A) reports the area per lipid, *a(x)* of these bilayers. It is observed that different bilayers have different *a(x)*, which implies that in the equimolar asymmetric bilayers, the leaflet with smaller *a(x)* is under tensile stress and the other one is under compressive stress. Therefore, to relieve these stresses, we constructed "relaxed" asymmetric bilayers by taking the number of lipids + cholesterol in each leaflet as the overall area of the bilayer divided by the *a(x)* of the leaflet (**Table A4**; Appendix A). Therefore, in these asymmetric bilayers, each leaflet now retained its *a(x)* equal to the value found in its symmetric bilayer. For these asymmetric bilayers, we did not find any significant transfer of cholesterols between the leaflets. Interestingly, this implies that cholesterol molecules transfer between the leaflets of equimolar asymmetric bilayers in order to relieve the stresses. Interestingly, we find that the *a(x)* of the leaflets of the equimolar asymmetric bilayers is same as that of symmetric bilayers of these lipids. An interesting question that arises from our results is whether the naturally occurring asymmetric bilayers are equimolar in lipid composition or have different number of lipids in the two leaflets so as to relieve stress. In prior work on synthetic asymmetric bilayers, the researchers have assumed equimolar lipid composition in the leaflets.[67] In this case, one would expect cholesterol to get re-distributed to relieve the stress.

To estimate the contribution of cholesterol and lipid molecules on the area per lipid of a leaflet, one can calculate the partial molar areas as suggested by Edholm and Nagle.[68] We have performed this calculation on the DPPC/CHOL and DOPC/CHOL symmetric bilayers. The partial molar areas of cholesterol and lipid, $a_{chol}(x)$ and $a_{lipid}(x)$, can then be estimated based on **Equation 4** and **Equation 5**:

$$a_i(\mathrm{x}) = \left(\frac{\partial A(N_1, \dots, N_m)}{\partial N_i}\right)_{N_{j\neq i}} \qquad \text{Equation 4}$$

$$a_{chol}(x) = a(x) + (1 - x)\frac{\partial a(x)}{\partial x} \quad \text{and} \quad a_{lipid}(x)$$
$$= a(x) - x\frac{\partial a(x)}{\partial x}$$

Equation 5

Figure 6 shows a plot of $a(x)$ versus x for DPPC/CHOL and DOPC/CHOL symmetric bilayers. It is observed that for the DOPC/CHOL bilayer, the $a_{chol}(x)$ and $a_{DOPC}(x)$ do not vary with x and are found to be $a_{chol}(x) = 0.21 \pm 0.01$ nm^2 and $a_{DOPC}(x) = 0.69 \pm 0.01$ nm^2. In the case of DPPC/CHOL bilayer, the partial molar areas depend on x and are listed in **Table A5** (Appendix A). In the case of DPPC/CHOL, the $a_{chol}(x)$ is close to zero for small values of cholesterol mole fraction. This result, which has been reported before[68], is due to the condensation effect wherein lipids in the vicinity of cholesterol molecules get more ordered resulting in only a small increase or even a decrease in the area. As the concentration of cholesterols increase, the effect of addition of one more cholesterol molecule is smaller and thus the $a_{chol}(x)$ increases. The S_{chain} of the DOPC/CHOL and DPPC/CHOL bilayers are listed in **Table A5** (Appendix A). It is observed that the S_{chain} in the case of DOPC/CHOL bilayers changes linearly from 0.23±0.02 to only 0.37±0.02 as the cholesterol amount increases from 0% to 50%. Therefore, the condensation effect in this case is weak and the partial molar areas of cholesterol and lipids remain constant as a function of cholesterol concentration. On the other hand, in the case of DPPC/CHOL bilayers, the S_{chain} increases non-linearly and at a more appreciable rate from 0.40±0.02 to 0.79±0.02 indicating that the condensation effect is strong. As a result, the partial molar areas of cholesterol and lipids vary with cholesterol concentration.

Figure 6

Area per lipid, $a(x)$, as a function of cholesterol mole fraction.
$x = N_{chol}/(N_{lipid} + N_{chol})$. A) DOPC/CHOL and B) DPPC/CHOL symmetric bilayers.
The $a(x)$ for the DOPC/CHOL bilayer varies linearly with x implying that the $a_{DOPC}(x)$
and $a_{chol}(x)$ are constants. On the other hand, the $a(x)$ for the DPPC/CHOL bilayer
shows some curvature implying that the partial molar areas are not constants for this
bilayer.

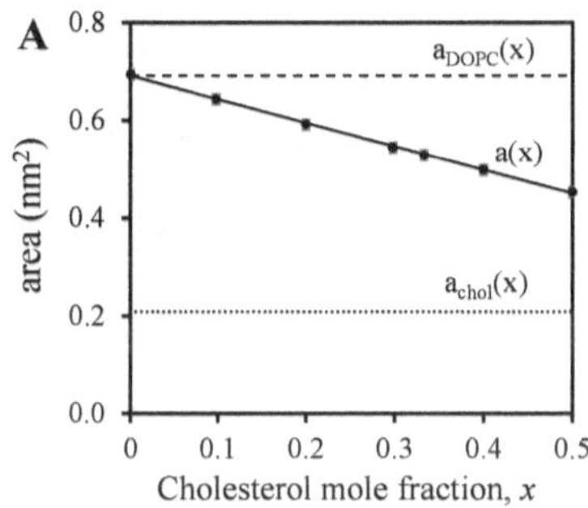

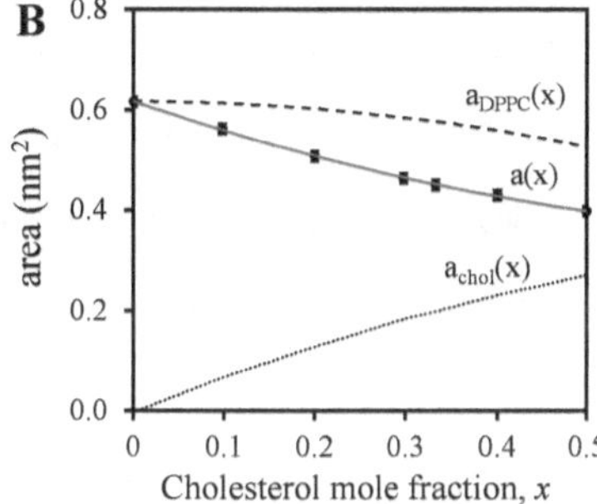

Conclusion

We have studied the distribution of cholesterol in various asymmetric bilayers. In

the bilayer mimicking the composition of the bilayer of red blood cells, we report that

cholesterol is enriched in the outer leaflet, which is comprised of more saturated

phospholipids, and which shows a higher lipids chain order than the inner leaflet. By

studying the distribution of cholesterol in several other equimolar asymmetric bilayers,

we show that the distribution of cholesterol is strongly coupled to the ordering of lipids in

the leaflets. In fact, our results demonstrate that the ratio of the distribution of cholesterol

in the leaflets is close to the ratio of the chain order parameter of lipids in the two leaflets.

This observation provides a quantitative relationship for predicting the distribution of cholesterol in equimolar asymmetric lipid bilayers. By simulating the lipid bilayers devoid of any cholesterol, we conclude that the ordering of the lipids in the leaflets governs the distribution of cholesterol in these bilayers. Ordering of lipids in the bilayers dictate the value of area per lipid. Since the area per lipid values for different lipid types is different, equimolar asymmetric bilayers are under mechanical stress: the leaflet with smaller area per lipid, that is higher lipid order, is under tensile stress and the other one (smaller lipid order) is under compressive stress. Therefore, transfer of cholesterol molecules occurs from the leaflet with compressive stress to the one with tensile stress. Interestingly, we find that the area per lipid of the leaflets of asymmetric bilayers remain the same as for the symmetric bilayers. Furthermore, our results show that the cholesterol flip-flop rate is also a function of ordering of lipids in the bilayer. The more ordered lipids are in the leaflets, the lower is the flip-flop rate. Cholesterol is understood to play an important role in modulating the physical and biological characteristics of lipid bilayers. Our findings highlight the strong relationship between cholesterol distribution and lipids ordering within asymmetric lipid bilayers and therefore provides useful insights in the field.

Chapter 3: Relationship Between Cholesterol Distribution and Phase of Bilayer

The material presented in this chapter is put together as a paper: "Aghaaminiha, M.; Farnoud, A. M.; Sharma, S. Distribution and orientation of cholesterol in leaflets and midplane of lipid bilayers as a function of membrane domains. *Journal of Physical Chemistry B* **2021**". [under review]

Introduction

Biological membranes, the semi-permeable interface that allows cells to interact with extracellular environment, are made up of proteins, lipids, sterols, and other molecules.[50,69] The lateral mobility and function of membrane proteins, is understood to be dictated by the structure and dynamics of the membrane.[70,71] Cholesterols, the main sterol in the membranes, affect the bending modulus,[72] and phase behavior of the membrane,[73] and are understood to play a role in the formation of rafts domains.[74] In addition, cholesterols are suspected to affect membrane fluidity and permeability,[75] lateral diffusivity, thickness, and structural orientation,[76,77] and lateral density of lipids in the membrane.[78] A lipid bilayer forms the dominant constituent of biological membranes, and so understanding their behavior as a function of cholesterol concentration will provide useful insights on the functioning of biological membranes.[79–81]

Solubility of cholesterol in a lipid bilayer is understood to be governed by the size of the head group of lipids and lipids' packing density.[82,83] The head groups of lipids shield cholesterols from water.[84] Therefore, cholesterols have higher solubility in lipids with a larger head group, such as sphingomyelins. Furthermore, it is reported that the solubility of cholesterol in membranes composed of saturated lipids is significantly higher in comparison to membranes with mono-/un-saturated lipids.[85]

Lipid bilayers are known to predominantly display two main domains: ordered and disordered. These domains can be distinguished by considering translational order and chain configurational order of lipids in the bilayers.[77,82] The translational order refers to local structure of lipids in the bilayer. Chain configurational order is a measure of the alignment of acyl chain of lipids in the bilayer. In ordered domains, both the translational order and the chain configurational order are high, while in disordered domains, both these orders are low. In binary/ternary mixtures, cholesterol is understood to favor the formation of the ordered domains by enhancing the chain configurational order.[37,86]

Previous studies have highlighted that there are numerous factors that govern the distribution of cholesterol molecules in lipid bilayers. However, no rational way is reported to predict the distribution as a function of composition and domains of the bilayers. In this work, we have employed molecular simulations to perform a systematic study of the relationship between distribution of cholesterol in the lipid bilayers as a function of cholesterol concentration and temperature. We show that the distribution of cholesterol molecules in the bilayer leaflets can be explained from the degree of order of the lipids in the bilayer. These simulations have been performed using the coarse-grained (CG) Dry Martini force field.[52] The CG Dry Martini force field treats water in an implicit manner. The absence of explicit water molecules makes simulations, computationally, much more efficient than when water molecules are treated explicitly in the simulation. However, an important concern is whether the results obtained from Dry Martini are comparable to the lipid force fields with explicit water. Therefore, before embarking on our simulations employing Dry Martini, we have performed a systematic comparison of the results obtained from an improved CG Wet Martini force field with a polarizable water model,[43] and those from Dry Martini. We show that the results obtained from these two force fields are comparable for the purpose of this study.

Computational Methods

Systems Setup

For comparing the results from Dry Martini and Wet Martini (study-I), we have studied the three types of lipid bilayers, namely, DOPC/CHOL (1, 2-dioleoyl-sn-glycero-3-phosphocholine/cholesterol), 16:0SM/CHOL (d18:1/16:0 sphingomyelin/Cholesterol), and DOPC/16:0SM/CHOL for a temperature range of 350 K to 290 K and cholesterol concentrations from 0% to 50% mole-ratio.

After establishing that the Dry Martini force field is suitable for studying properties of lipid bilayers at different cholesterol concentrations, we have examined the spatial distribution of cholesterol molecules in DOPC/CHOL, 20:0SM/CHOL (d20:1/20:0 sphingomyelin /Cholesterol), and DOPC/20:0SM/CHOL lipid bilayers over a wider range of temperatures and cholesterol concentrations using the Dry Martini force field (study-II). The temperature and concentration of cholesterol in study-II was varied from 210 K to 400 K and 0% to 50% mole-ratio, respectively. We have chosen temperatures beyond the range that is biologically relevant so as to clearly identify phase transitions in the lipids. For example, the melting temperature of DOPC is quite low ($\sim$ 253 K)[87] while that of 20:0SM is relatively high ($\sim$ 319 K).[88]

Simulations Parameters

In all the simulations, periodic boundary conditions are applied in the three directions of the simulation box. In both study-I and study-II, the simulation box size is 125 Å in all directions. The total number of lipids (phospholipids + cholesterol) for all systems is kept constant to be 510±2. In the simulations where CG Wet Martini force field is employed, the total number of water molecules is 10,850±40. **Figures B1** (Appendix B) displays the topologies of lipids together with the name and type of each bead. Initial configurations of the bilayer systems are constructed using the *insane.py* script developed by Marrink et al.[45] The simulations are performed using GROMACS/5.1.2 molecular simulation package.[46]

The initial configuration is energy minimized using steepest descent with the maximum force tolerance of 1000.0 KJ/mole/nm. To pre-equilibrate the bilayer structure, a 25 ns canonical ensemble (constant number of particles N, constant volume V and constant temperature T) simulation followed by a 25 ns isothermal-isobaric ensemble (constant N, pressure P, and T) simulation is performed at each temperature. The time-step for the pre-equilibration simulations is 10 fs. After pre-equilibration, a production run of 1 μs is performed at each temperature in the NPT ensemble and the data was collected for the last 500 ns of each simulation.

For the systems where CG Dry Martini force field is used, simulation parameters are chosen same as the Arnarez et al. study that parametrized the implicit water model on the original Martini force field.[52] However, simulation parameters of systems with Wet Martini force field are taken from the work of Michalowsky et al., where the polarizable water model of the Martini force field is parameterized.[43] **Table B1** (Appendix B) summarizes the simulation parameters used in all the coarse grained simulations of lipid bilayers studied in this research.

As mentioned earlier, in study-I, temperature range is taken from 350 K to 290 K, and in study-II, it is 400 K to 210 K. The temperature decrement step in both studies is 10 K. At each temperature, for all the simulations, a short 50 ns pre-equilibration followed by 1 µs production run is accomplished with the parameters presented in **Table B1** (Appendix B) and the data is collected for the last 500 ns of the production run.

Lipid Bilayers Properties

Area per lipid (*apl*) is calculated as the ensemble-averaged area of the plane of the simulation box in which the bilayer resides, divided by the total number of two-chain lipids + cholesterols in one leaflet.[51] Bilayer thickness (D_{P-P}) is calculated as the ensemble-averaged distance between the phosphate (PO_4 bead; shown in **Figure B1**; Appendix B) beads of lipids in each leaflet.[52] The orientation of a cholesterol molecule within a bilayer is defined as the angle between vector, $\bar{a}$ (the vector joining the ROH bead to the C1 bead of cholesterol; see **Figure B1**; Appendix B) and vector $\bar{z}$ (the bilayer normal). To quantify chain configurational order of lipids, we calculate the ensemble-averaged order parameter of the acyl chain of lipids (S_{chain}) according to **Equation 6**:[2,50]

$$S_{Chain} = \langle \frac{1}{2}(3\langle cos^2\theta \rangle - 1) \rangle \qquad \text{Equation 6}$$

Where θ is the angle between each bond vector of the lipid beads and the bilayer normal (*z*-axis). A large value of the S_{chain} (~ 1) indicates that the bonds are aligned with the bilayer normal, representing the high chain configurational order characteristic of the ordered domains,[2,50] while its small value (~ 0) indicates that the bonds are randomly oriented with respect to the bilayer normal, a signature of the disordered domains.[2,50] To quantify translational ordering of lipids, 2D radial distribution functions ($g_{2D}(r)$) are calculated. We calculate the $g_{2D}(r)$ based on the C1B beads (see **Figure B1**; Appendix B) of lipids, representing the center of mass of lipids, as suggested elsewhere.[50]

Results and Discussion

Comparison of Dry Martini and Wet Martini force fields – Study I

To compare the performance of Dry Martini and Wet Martini force fields in simulating lipid bilayers behavior under different conditions, we studied three systems presented in **Table 3**. First, we discuss the comparison of experimental results with our results for the pure DOPC and pure 16:0SM bilayers. Then, we compare the Dry and Wet Martini force fields in the lipid bilayers with a cholesterol concentration in the range of 10% to 50% mole-ratio.

Table 3

Studied lipid bilayers to compare Dry and Wet Martini force fields.
In the ternary mixtures, always, DOPC and 16:0SM exist equally in the systems (i.e., 1:1 mole-ratio).

Lipid bilayer	# Simulations	Temperature range	Cholesterol mole-ratio range
DOPC/CHOL	$7 \times 6 = 42$	350 K to 290 K	0% to 50%
16:0SM/CHOL	$7 \times 6 = 42$	350 K to 290 K	0% to 50%
DOPC/16:0SM/CHOL	$7 \times 6 = 42$	350 K to 290 K	0% to 50%

First, we compare the Dry and the Wet Martini force fields in simulations where cholesterol is absent in the bilayer and experimentally/theoretical results are also available for those systems. *Pure DOPC bilayer*: In an experimental study by Petrache et al. in 2004 the values of area per lipid and bilayer thickness of pure DOPC system are reported to be 0.725 nm^2 and 36.9 Å, respectively.[89] Liu and Qi in 2012 studied DOPC lipid bilayers via MD simulations at 303 K and reported 0.691 ± 0.013 nm^2 and 37.47 ± 0.37 Å as area per lipid and bilayer thickness, respectively.[90] Also, from the simulations of fully atomic systems for pure DOCP bilayer, the area per lipid is reported to be 0.705 – 0.711 nm^2 and the bilayer thickness as between 35.08 – 35.80 Å.[91,92] Comparing our results (presented in **Table B2**; Appendix B) with the results from these experimental and theoretical studies for DOCP bilayer reveals that the Dry Martini force field matches the experimental value of bilayer thickness quite well (~ 0.6% error). Furthermore, Wet Martini force filed did better estimation for area per lipid (~ 5.6% error). In overall, the results from Dry and Wet Martini are in good agreement with each other and with the experimental and computational results for the pure DOPC bilayer. *Pure 16:0SM bilayer*: Li et al. studied the interfacial behavior of different acyl chain length sphingomyelin molecules in experiments. They reported the average area per lipid for 16:0 and 18:0SM lipid bilayers as 0.525 and 0.486 nm^2, respectively at 303 K.[88] Maulik et al. reported a slightly smaller value 0.47 nm^2 for area per lipid of 16:0SM at 328 K.[93] From fully atomistic MD simulations, the area per lipid of 16:0SM is found to be 0.52 ± 0.01 nm^2 at 323 K.[94] As discussed in a previous work[58], there are some difficulties in assessing area per lipid experimentally, which may explain the variation in the experimentally reported results of 16:0SM. The bilayer thickness (D_{P-P}) reported for 16:0SM at 323 K, from experiments and MD simulation studies, are 44.40 Å[93] and 43.40 Å[94], respectively. Our results for 16:0SM bilayer (presented in **Table B2**; Appendix B), agree well with the experimental and computational results.

Next, we compare the Dry and the Wet Martini force fields in the simulations where cholesterol is present in the bilayer. As shown in **Table 3**, we have analyzed several systems for comparing the Dry and the Wet Martini force fields. **Figure 7** shows the comparison of results obtained from the Dry and the Wet Martini force fields for area per lipid, bilayer thickness, chain order parameter, and 2D radial distribution functions of DOPC/16:0SM/CHOL bilayer with 30% mole-ratio of cholesterol. Same comparison of Dry martini and Wet Martini force fields for DOPC/CHOL and 16:0SM/CHOL bilayers with 30% mole-ratio of cholesterol is reported in **Figure B2** and **Figure B3** (Appendix B), respectively.

Figure 7

*Comparison of the Dry and Wet Martini force fields
in investigation of structural properties of DOPC/16:0SM/CHOL lipid bilayer at
temperature range of 350 K to 290 K, and cholesterol concentration of 30% mole-ratio.
A) Area per lipid, apl; B) Bilayer thickness, DP-P; C) Chain order parameter, Schain;
D) 2D RDFs of C1B-C1B beads of two-chain lipids for simulations performed with Dry
Martini force field; and E) 2D RDFs of C1B-C1B beads of two-chain lipids for
simulations performed with Wet Martini force field. Successive g2D(r) have been shifted
vertically by 0.4 units for the sake of clarity.*

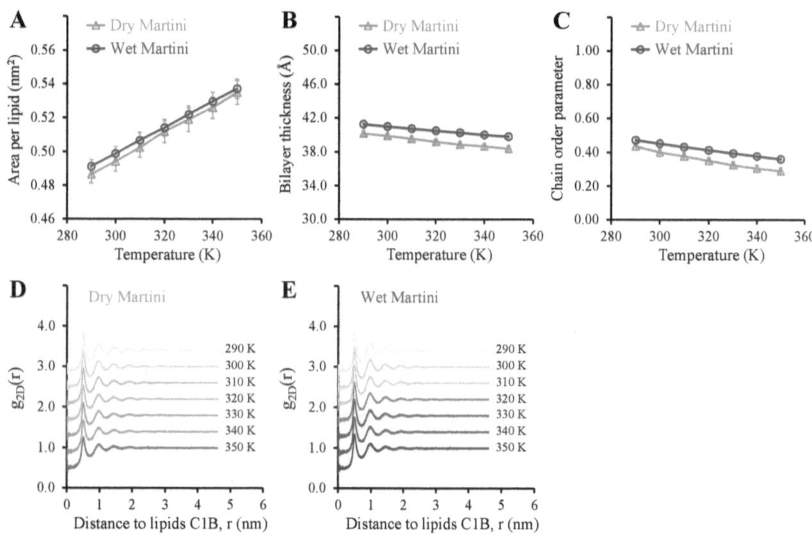

Cholesterol Distribution as a Function of Lipid Bilayers Domains – study II

Study-1 shows that the results from Dry Martini are comparable to those from

Wet Martini. Now, we employ the Dry Martini force field to perform a systematic

investigation of how the domain/phase characteristics of lipid bilayers affect the

distribution of cholesterol molecules within the bilayer. For this purpose, we have studied three two-chain lipids + cholesterol mixtures presented in **Table 4**. First, we show how the phase transition temperature, *Tm*, of lipid bilayers is calculated. Then, the effect of cholesterol concentration on the *Tm* is discussed, and finally the relationship between *Tm* and cholesterol localization and orientation within the bilayers is highlighted.

Table 4

Studied lipid bilayers to investigate the relationship between domain/phase of lipid bilayers and cholesterol localization within the membrane.
In the ternary mixtures, always, DOPC and 20:0SM exist equally in the systems (i.e., 1:1 mole-ratio).

Lipid bilayer	# Simulations	Temperature range	Cholesterol mole-ratio range
DOPC/CHOL	$20 \times 6 = 120$	400 K to 210 K	0% to 50%
20:0SM/CHOL	$20 \times 6 = 120$	400 K to 210 K	0% to 50%
DOPC/20:0SM/CHOL	$20 \times 6 = 120$	400 K to 210 K	0% to 50%

***Tm* of pure lipid bilayers.** We have calculated structural properties and 2D RDFs of lipid bilayers of pure DOPC, pure 20:0SM and 1:1 mixture of DOPC/20:0SM in the absence of any cholesterol. **Figure 8** shows area per lipid, *apl*, average bilayer thickness, $D_{P\text{-}P}$, average chain order parameter, S_{chain}, and 2D radial distribution function, $g_{2D}(r)$, of C1B-C1B beads of two-chain lipids. From the structural properties of lipid bilayers presented in **Figure 8**, it is observed that 20:0SM and 1:1 mixture of DOPC/20:0SM have significant ordering of lipids at low temperatures, and as the temperature increases, they undergo an ordered $\rightarrow$ disordered transition. In contrast, pure DOPC bilayer does not show significant ordering even at the lowest temperatures studied. The transition temperature, Tm, is estimated by plotting the derivative of *apl*, $D_{R\text{-}R}$, and S_{chain} of each system and then setting the second derivative to zero. **Figure B4** (Appendix B) shows a plot of the derivatives of all these structural properties. The *Tm* of pure 20:0SM is estimated to be 310 K, which agrees well with the experimentally reported value of 310 − 320 K.[88,94] The Tm of 1:1 mixture of DOPC/20:0SM system is estimated to be 250 K. For pure DOPC, the first derivative is zero at 265 K. The Tm of pure DOPC estimated in this manner agrees well with the experimental value of 255 ~ 260 K.[87] For both 20:0SM and DOPC/20:0SM systems, the $g_{2D}(r)$ are shown in **Figure 8** (**E** and **F**), respectively. The $g_{2D}(r)$ show loss of long-range structure as the temperature increases beyond the *Tm*. The $g_{2D}(r)$ of DOPC do not show any long-range order. The behavior of $g_{2D}(r)$ along with other structural properties of DOPC lipid bilayer, shown in **Figure 8**, implies that this bilayer does not have a strong translational ordered structure even at low temperatures. Another important observation from **Figure 8** (**A** to **C**) is that the binary mixture of DOPC/20:0SM system is always in between its two single-lipid bilayers. This statement means that at low temperatures the binary system behaves more like the pure 20:0SM while at high temperatures it behaves more like DOPC. This observation suggests that in the binary system of DOPC/20:0SM, at low temperatures, the behavior of 20:0SM, a saturated phospholipid with long tail and sphingosine backbone, is dominant in the bilayer while at high temperatures, DOPC which is an unsaturated two-chain phospholipid with a glycerol backbone, dominates the behavior of the binary mixture.

Figure 8

Phase transition temperature estimation for bilayers with no cholesterol.
The Tm values are estimated according to the change in structural properties. A) Area
per lipid, apl; B) Bilayer thickness, DP-P; C) chain configurational order, Schain; D) 2D
RDFs of C1B-C1B beads in pure DOPC bilayer; E) 2D RDFs of C1B-C1B beads in pure
20:0SM bilayer; and F) 2D RDFs of C1B-C1B beads in 1:1 DOPC/20:0SM bilayer.
Successive g2D(r) have been shifted vertically by 0.4 units for the sake of clarity.

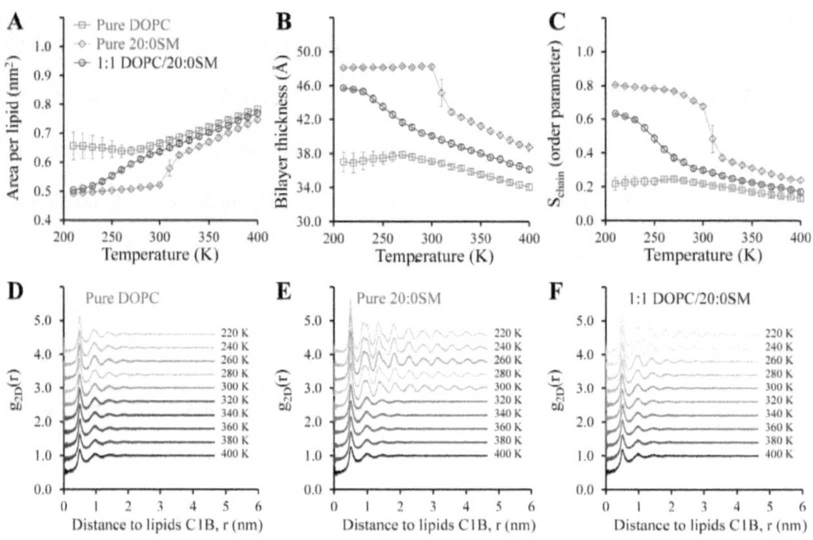

***Tm* versus Cholesterol Concentration.** In the next step, the effect of cholesterol

concentration on the structural properties of lipid bilayers and as a result phase transition

temperature, *Tm*, is studied. As explained earlier, the *Tm* is calculated by plotting the

derivative of structural properties of each system and then setting the second derivative to

zero as well as analysis of the translational order by calculating $g_{2D}(r)$. **Figure 9** shows

how the structural properties of 20:0SM/CHOL bilayer change with the cholesterol concentration. Same analysis for DOPC/CHOL and DOPC/20:0SM/CHOL systems are presented in **Figure B5** (Appendix B). As discussed above, the *Tm* is estimated as the temperature at which the second derivative of structural properties (S_{chain} here) is zero. Thus, we have calculated the derivative of S_{chain} for DOPC/CHOL, 20:0SM/CHOL and DOPC/20:0SM/CHOL bilayers and have reported the results in **Figure B6** (Appendix B). It is observed that as the cholesterol concentration increases, the T_m shifts towards higher values. This shift in *Tm* with an increase in cholesterol concentration agrees with the reported results of DPPC/CHOL bilayer.[70] It is important to note that for the DOPC/CHOL bilayers (for all 10% to 50% mole-ratios of cholesterol) the derivative plots have no local maximum, and so *Tm* values of these bilayers are not reported. The calculated *Tm* values of all studied systems are presented in **Table 5**.

From **Figure 9A**, we can also observe that at all temperatures, systems with higher cholesterol concentration have smaller area per lipid, which agrees with the condensation effect of cholesterol. The condensation effect is a result of strong interactions that cholesterol has with amide and glycerol moieties of phospholipids.[25,56,95] **Figure 9B** shows that at high temperatures bilayers with higher cholesterol concentration are thicker as expected, however, at low temperatures, it is not the case. From **Figure 9B** we observe that at low temperatures, T < 300 K, systems with less amount of cholesterol are thicker. Same behavior for the chain configurational order parameters is also observed in **Figure 9C**. These results at low temperatures are interesting to our knowledge as, experimentally, at low temperatures the effect of increase in cholesterol concentration on lipid bilayer thickness have not been studied yet as most of

experimental works are performed above *Tm* of pure lipids present in the system.

Figure 9

Structural properties of 20:0SM/CHOL bilayers
as a function of temperature and cholesterol concentration.
A) Area per lipid, apl; B) Bilayer thickness, DP-P; and C) chain order parameter,
Schain. Error bars in all figures were estimated as standard error of the mean when the
last 0.5 µs of each system were split into three equally sized blocks and analyzed
separately. Same analysis for the DOPC/CHOL and DOPC/20:0SM/CHOL bilayers are
performed, and the results are reported in Figures B5 (Appendix B).

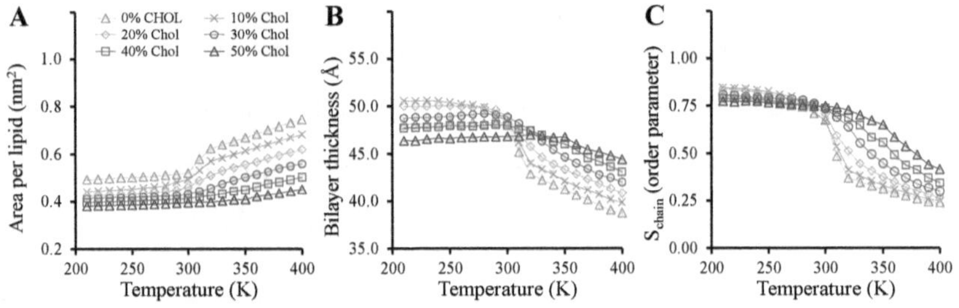

Table 5

Phase transition temperature estimation for lipid bilayer with cholesterol.
The Tm values are calculated by plotting the derivative of apl, DR-R, and Schain of each
system and then setting the second derivative to zero. For the DOPC/CHOL bilayers the
derivative plots had no relative maximum, thus, no values for these bilayers are reported.

Lipid bilayer	0% Chol	10% Chol	20% Chol	30% Chol	40% Chol	50% Chol
DOPC/CHOL	270 K	NA	NA	NA	NA	NA
20:0SM/CHOL	310 K	310 K	310 K	330 K	350 K	370 K
DOPC/20:0SM/CHOL	250 K	260 K	270 K	290 K	310 K	335 K

Cholesterol Distribution in Lipid Bilayers. In our previous research work we have studied the distribution of cholesterol in various asymmetric lipid bilayers.[14,96] In that study, we have shown that cholesterol distributes between the inner and the outer leaflets of lipid bilayers so as to relieve tensile and compressive mechanical stresses between the two leaflets. In lipid bilayers with the same number of two-chain lipids in the inner and outer leaflets, the leaflet with more ordered lipids is in tensile stress while the other leaflet is in compressive stress. So, cholesterol molecules concentrate in the leaflet with the tensile stress to relieve this difference of stresses between the leaflets. In the current research, we have attempted to relate the distribution of cholesterol with the phase of the bilayer in different symmetric lipid bilayers. The main two phases, *ordered* vs *disordered*, are related to the angle and location of cholesterol molecules within the bilayers. To quantify this relationship, we have divided the bilayer into two regions of midplane and leaflets. The threshold for separating these regions is set to be nearly one third of the bilayers' thickness with the values of 13Å, 14Å, and 15Å for DOPC/CHOL, DOPC/20:0SM/CHOL, and 20:0SM/CHOL bilayers, respectively. Cholesterol molecules located beyond the threshold from the center plane of the bilayer are considered to be inside the leaflets, while cholesterol molecules within the threshold from the center plane are counted to be in the midplane region. This analysis agrees with the previously reported threshold of 15 Å for DPPC/CHOL lipid bilayers with 10% to 50% mole-ratios of cholesterol at 300 K to 350 K.[49]

Figure 10 shows the orientation of cholesterol molecules at different locations of the three studied lipid bilayers named in **Table 4** with 30% mole-ratio of cholesterol at three different temperatures, one above the phase transition temperature (*Tm*), one close to the *Tm*, and one below the *Tm*. **Figure B7** (Appendix B) displays the same analysis for the studied systems with other mole-ratios of cholesterol from 10% to 50%. For this purpose, we have divided the bilayers into 50 slabs parallel to the *xy* plane. In each slab, the ensemble-averaged angle of cholesterol molecules with the z axis (bilayer normal) is calculated. The reported angles correspond to the average over configurations of three equally sized blocks of simulations in the last 0.5 μs of each system. According to **Figure 10**, in all three systems, cholesterol molecules located in the leaflets orient with bilayer normal with angle $\theta < 30°$ or $\theta > 150°$, while, at the midplane $30° < \theta < 50°$. An important observation from **Figure 10**is that the cholesterol orientation changes almost linearly with the distance from the center of the bilayer for temperatures higher than the *Tm* for all systems. However, when the temperature is lower than the *Tm*, this linear behavior is still valid for the DOPC/CHOL bilayer, but it is not the case for the 20:0SM/CHOL and DOPC/20:0SM/CHOL bilayers. For these two bilayers, **Figure 10 (B** and **C)**, the cholesterol molecules orient relatively parallel to the bilayer normal even at the midplane region.

Figure 10

Cholesterol orientation with respect to its location in the bilayer.
Ensemble-averaged angle of cholesterol molecule with the bilayer normal as a function
of the location of the hydroxyl headgroup of cholesterol in 30% mole-ratio of lipid
bilayers. A) DOPC/CHOL, B) 20:0SM/CHOL, and C) DOPC/20:0SM/CHOL. Shaded
area represents the midplane of the bilayer (13Å, 14Å, and 15Å from the center of the
bilayers in figures A, B, and C, respectively). Error bars in all figures were estimated as
standard error of the mean when the last 0.5 μs of each system were split into three
equally sized blocks and analyzed separately. Complete analysis for the systems of
DOPC/CHOL, 20:0SM/CHOL, and DOPC/20:0SM/CHOL with other mole-ratios of
cholesterol are performed, and the results are reported in Figures B7 (Appendix B),
respectively.

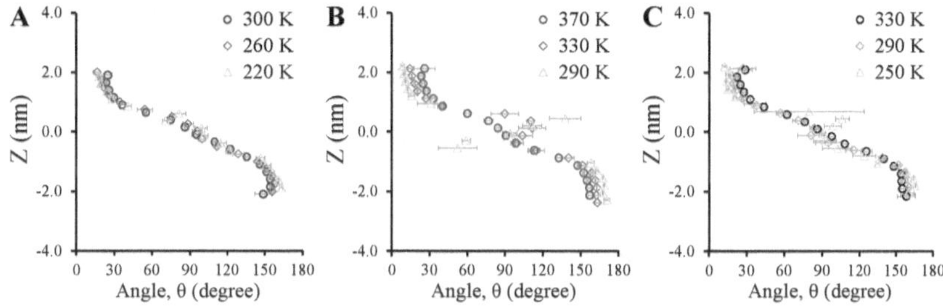

Cholesterol Location vs Cholesterol Concentration. To examine how the

location of cholesterol molecules varies with mole-ratio of cholesterol, we have done

further analysis for the mole-ratios of cholesterol varying from 10% to 50% for all three

lipid bilayers presented in **Table 4**. As discussed before, we have divided the bilayer into

two regions of leaflets and midplane. **Figure 11** displays fraction of cholesterol

molecules in the two regions of the three studied bilayers where mole-ratio of cholesterol

is 30%. For all the systems presented in **Figure 11,** it has been observed that there is less

fraction of cholesterol molecules in the leaflets at higher temperatures. As the

temperature decreases, the fraction of cholesterol molecules in the midplane region reduces and at a certain temperature a crossover in the fraction of cholesterol in the leaflets and midplane of the bilayers is observed. This temperature is termed as *Tcross*. Complete analysis for the systems of DOPC/CHOL, 20:0SM/CHOL, and DOPC/20:0SM/CHOL are performed, and the results are reported in **Figures B8** (Appendix B). From these figures, we can observe that the *Tcross* is positively correlated with cholesterol concentration. In other words, for the bilayers with higher mole-ratio of cholesterol, the crossover happens at the higher temperature while for the bilayers with lower cholesterol concentration, the crossover happens at lower temperatures. For the DOPC/CHOL system with less than 20% mole-ratio of cholesterol (**Figure B8 – A1** and **A2**; Appendix B) no crossover is observed, which means in the binary bilayer of DOPC/CHOL for low concentration of cholesterol (< 20%), cholesterol molecules always prefer to be in the midplane region. The estimated *Tcross* values of all bilayers with different mole-ratio of cholesterol are reported in **Table 6**.

Table 6

Crossover temperature, Tcross, of studied lipid bilayers.
Tcross values are temperatures at which the fraction of cholesterol in the leaflets and the midplane of the bilayer becomes equal. Figures B8 (Appendix B) display how the Tcross values are estimated.

Lipid bilayer	0% Chol	10% Chol	20% Chol	30% Chol	40% Chol	50% Chol
DOPC/CHOL	NA	–	–	255 K	295 K	330 K
20:0SM/CHOL	NA	300 K	315 K	330 K	345 K	370 K
DOPC/20:0SM/CHOL	NA	255 K	260 K	290 K	315 K	345 K

Figure 11

Fraction of cholesterol molecules at different regions of bilayers.
A) DOPC/CHOL, B) 20:0SM/CHOL, and C) DOPC/20:0SM/CHOL lipid bilayer for 30%
mole-ratio of cholesterol. Error bars in all figures were estimated as standard error of
the mean when the last 0.5 μs of each system were split into three equally sized blocks
and analyzed separately. Estimated Tcross values are 255±5 K, 330±5 K, and 285±5 K
for figure A, B, and C, respectively. Complete analysis for the systems of DOPC/CHOL,
20:0SM/CHOL, and DOPC/20:0SM/CHOL are performed, and the results are reported
in Figures B8 (Appendix B). Also, Table 6 summarizes the Tcross values of all studied
bilayers.

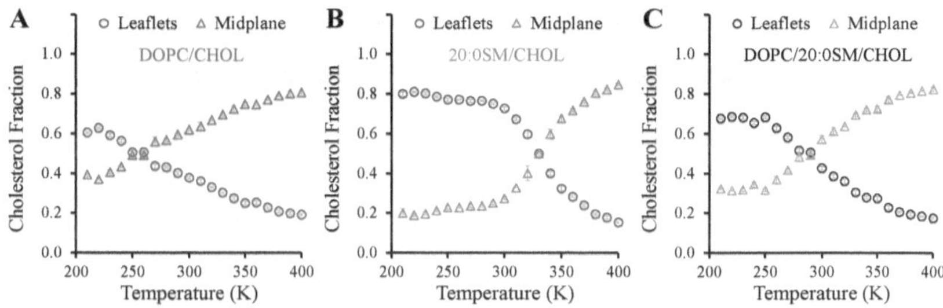

Cholesterol location vs phase of the bilayer. To examine if the location of

cholesterol molecules is meaningfully related to the phase of the bilayer, in **Figure 12**,

we have plotted the *Tm*, against *Tcross*. From this figure we can clearly observe that for

the same mole-ratio of cholesterol, the *Tm* and *Tcross* values of 20:0SM/CHOL and

DOPC/20:0SM/CHOL lipid bilayers are strongly correlated. This positive correlation

means that cholesterol location within the bilayer is a function of the phase of the bilayer.

As we observed in **Figure 10**, cholesterol molecules at the midplane orient to the bilayer

normal with $30° < \theta < 130°$ while at the leaflets $\theta < 30°$ or $\theta > 150°$. **Figure 11** also shows

that at $T > Tcross$, cholesterol molecules are majorly located at the midplane of the

bilayers while at $T < Tcross$ the leaflets are more favorable for cholesterol molecules to be occupied in. **Figure 12** additionally displays that the *Tm* ~ *Tcross*. Summing these statements together, we can conclude that when the bilayer is in the *disordered* phase, $T >$ *Tm*, cholesterol molecules are dominantly distributed at the midplane of the bilayer and orient to the bilayer normal with the angle of $30° < \theta < 150°$, while at the *ordered* phase, $T < Tm$, cholesterol molecules are dominantly distributed at the leaflets of the bilayer and orient to the bilayer normal with the angle of $\theta < 30°$ or $\theta > 150°$. It is important to mention that in this study we have not discussed the difference of inner and outer leaflets in terms of cholesterol distribution, however, we have done analysis of cholesterol distribution in complex asymmetric bilayers mimicking composition of red blood cells' membrane in our previous work[14,97,98] and have reported that the cholesterol molecules are asymmetrically distribute in asymmetric bilayers mainly to relieve the difference of stress between the leaflets.

Figure 12

Tm versus Tcross of studied lipid bilayers.
Relationship between phase of lipid bilayers in terms of Tm and location of cholesterol
molecules within the bilayers in term of Tcross. For DOPC/CHOL bilayers, the Tm plot
is not shown as no phase transition temperature is reported for these systems in Table 5.
Also, for this system, no crossover for the bilayers with cholesterol mole-ratios > 30% is
found according to Table 6 of this study. The error bars are approximated to be 10 K.

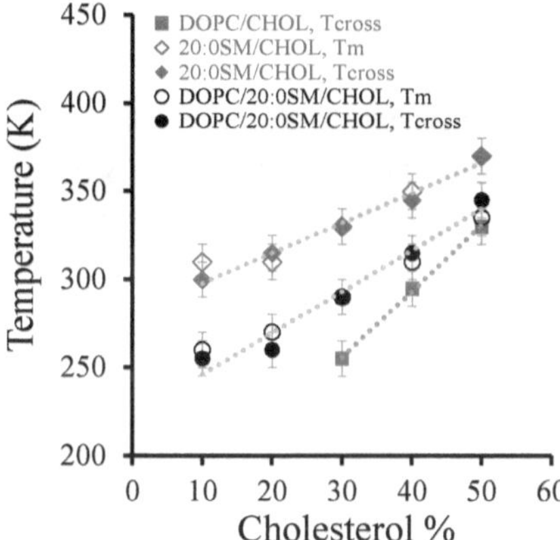

Conclusion

- Dry Martini force field can capture structural properties of lipid bilayers same as
  Wet Martini force filed. Area per lipid, bilayer thickness, and chain order
  parameter of our examined lipid bilayers with no cholesterol are calculated and
  compared with empirical data, and results are in well agreement.

- We have studied spatial distribution of cholesterol molecules as a function of
  cholesterol concentration and temperature in three lipid bilayer systems:
  DOPC/CHOL, 20:0 SM/CHOL and DOPC/20:0 SM/CHOL using coarse-grained
  molecular simulations. We find that the spatial distribution of cholesterol
  molecules is closely related to the *ordered* and disordered domains of the bilayers.

In the *ordered* domains, cholesterol molecules prefer to be in the leaflets and align themselves parallel to the bilayer normal. In the disordered domains, cholesterol molecules are preferentially in the midplane region of the bilayer.

- Although it is known that cholesterol is enriched in the ordered domains of plasma membrane, what we observed is that in the ordered domains cholesterol is located highly aligned in the leaflets with the angle $\theta<30°$ or $\theta>50°$.

- At disordered domains of lipid bilayers, cholesterol molecules are located at the midplane of the lipid bilayer with $30°<\theta<150°$.

- Distribution of cholesterol molecules in lipid bilayers can be viewed as a useful parameter to detect if ordered domain is dominant in the bilayer or it is vice versa.

Chapter 4: Estimation of Phase Diagrams for Three-Component Lipid Mixtures

The material presented in this chapter is put together as a paper: "Aghaaminiha, M.; Ghanadian, S. A.; Ahmadi, E.; Farnoud, A. M. A Machine Learning Approach to Estimation of Phase Diagrams for Three-Component Lipid Mixtures. *Biochim. Biophys. Acta BBA - Biomembr* **2020**, *1862* (9), 183350".[99]

Introduction

Over the past few decades, significant evidence has gathered regarding the presence of lateral heterogeneities, commonly known as lipid domains or rafts, in the plasma membrane of eukaryotic cells. These phase separations are the result of the preferential aggregation of cholesterol and saturated, long-chain lipids.[73,86,100–102] Ordered lipid domains have been postulated to play an important role in a variety of cell functions including signal transduction,[86] endocytosis,[103] and cell division,[104] among others. Given the importance of lipid domains in various biological phenomena, significant effort has been focused on fundamental studies to examine lipid phase segregation.[79,80,84,105] However, due to the complexity of the plasma membrane, such studies are generally performed in well-defined model systems to provide information regarding the role of lipid chemical structure and molar ratio in the lateral heterogeneity of the membrane.

Phase diagrams have emerged as an important tool in characterizing lipid phase behavior in model membranes. First proposed by Feigenson and Buboltz,[84] such diagrams are generally developed for systems composed of three lipids: one lipid with a high melting temperature, a sterol (generally cholesterol), and another lipid with a low melting temperature.[73,101] The melting temperature (Tm) itself is highly dependent on the lipid chemical structure, such as chain length, headgroup composition, and number and type of unsaturation.[101,106] Examination of changes in the phase behavior of three-component lipid mixtures has led to the development of diagrams, in which various lipid phases, including liquid-disordered (Ld), liquid-ordered (Lo), and gel ($L\beta$) phase, as well as their coexistence, can be observed as a function of the composition of lipids, their molar ratio, and the temperature. Such phase diagrams have been instrumental in allowing researchers to draw generalizable conclusions on the role of each lipid component in lipid phase behavior.

While three-component phase diagrams are valuable in understanding and interpreting lipid phase behavior, they are difficult to develop experimentally. The development of phase diagrams generally requires more than one experimental technique and a combination of fluorescence spectroscopy, confocal microscopy, and sometimes X-ray based techniques are used.[79,80,105] Besides, determining the exact position of the tie-lines requires that a large number of samples be tested with significant precision to cover the entire composition space. For example, it has been estimated that approximately 400 samples of different compositions are needed to achieve five mole percent resolution for a three-component phase diagram.[80] More importantly, a simple change in the lipid composition of one of the components or changes in temperature would require repeating the experiments to redevelop the phase diagram for the new conditions. New techniques that could develop such phase diagrams without the need for expensive and time-consuming experimental procedures would significantly enhance the current understanding of lipid phase behavior without significant experimental costs.

Machine learning is a technique for mining dataset and using patterns and inference to develop new information. Supervised machine learning uses example input-output pairs to train an algorithm. In this case, the dataset is usually split into "training", "cross-validation", and "testing" categories. The training dataset is used to train a predictive model, validating dataset helps improve the model during training, and the testing dataset is used to evaluate the performance of the constructed predictive model. Machine learning has already found applications in the characterization of lipid-protein interactions in biological membranes.[107,108] However, the application of machine learning in analysis of membrane phase separation is only starting to gain attention, with two very recent reports focusing on the use of machine learning to analyze lipid phase separations based on results from molecular dynamics simulations[109] and to examine the role of lipid domains in the recruitment of proteins to transmembrane receptors.[110] However, despite the importance of phase diagrams in understanding the biophysical behavior of membrane lipids, and the availability of experimental information that can be used to train algorithms, the use of machine learning to develop phase diagrams has remained unexplored.

In the current study, we applied supervised machine learning to predict the melting temperature of phospholipids depending on their chemical structure and to develop three-component lipid phase diagrams for various lipid mixtures. First, the artificial neural network (ANN) was used for the prediction of phospholipid melting temperature. Next, random forests (RF) and support vector machines (SVM) were used to develop a three-component phase diagram for lipid mixtures containing cholesterol, a low

Tm, and a high *Tm* phospholipid. Results show, for the first time, that it is possible to utilize machine learning techniques to determine phospholipid melting temperatures that closely match values reported in the literature. Results also reveal that it is possible to use machine learning to develop three-component phase diagrams for various lipid mixtures, or the same mixture at different temperatures, without the need for experimentation. This work can pave the way for the application of machine learning techniques in the evaluation of lipid phase behavior, thereby helping save significant time and resources.

Material and Methods

Datasets

All three machine learning methods require training datasets. The training dataset for the ANN included the melting temperature of lipids reported in the literature, while the training dataset for the RF and SVM methods included the reported phase diagrams for known lipid mixtures, also acquired from literature reports.

Melting Temperature Dataset. The melting temperature dataset included 65 samples, all retrieved from the Lipid Thermotropic Phase Transition Database (LIPIDAT) – NIST Standard Reference Database 34.[111,112] Each data point was described by eleven features (or input variables), with *Tm* as the response variable (**Table C1**; Appendix C). Of the available *Tm* dataset, 70% of the data was used as the training set, 15% as cross-validation, and the rest (15%) as a testing set. These ratios were selected based on common practices in the machine learning literature. Dataset features included: *T1L* (tail 1 length, varied between 3 and 24), *T2L* (tail 2 length, varied between 3 and 24), *MW* (molecular weight in g/mole), *BT* (backbone type, glycerol: 1, sphingosine: 2), *HS* (head size, choline: 104 g/mole, ethanolamine: 61 g/mole, glycerol: 92 g/mole, serine: 105 g/mole, hydroxyl: 17 g/mole), *ChT* (acyl chain type, not mixed acyl chain: 0, mixed acyl chain: 1), *ST* (saturation type, saturated: 0, unsaturated-cis: 1, unsaturated-trans: 2), *NU* (number of unsaturated carbons, varied between 0 and 12), *HCh* (head charge, zwitterionic: 0, anionic: -1), *ID1* (first unsaturated carbon ID number on tail 1), and *ID2* (first unsaturated carbon ID number on tail 2). The response variable was *Tm*, the melting temperature of the lipid. Taking 16:0-22:6 phosphatidylcholine as an example, the features were as follow: *T1L:* 16, *T2L:* 22, *MW:* 806, *BT:* 1, *HS:* 104, *ChT:* 1, *ST:* 1, *NU:* 6, *HCh:* 0, *ID1:* 0, *ID2:* 4, and *Tm:* -27.

Phase Diagram Dataset. The phase diagram dataset was developed based on the three-component phase diagrams listed in **Figure 13** (see **Figure C1**, Appendix C, for more details). The first seven diagrams (**Figure 13A** to **G**) were used as the training and cross-validation dataset. The last diagram (H) formed the testing dataset. Each diagram was redrawn from the literature as a combination of 5,151 data points (see **Figure C2**; Appendix C). Therefore, the training dataset contained $7 \times 5,151 = 36,057$ samples. An abbreviated and not randomized version of the constructed training dataset is reported in **Table C2** (Appendix C). Each sample in the dataset was described by fifteen features and a response variable (class) as follows (note that Lipid 1 refers to the low *Tm* lipid and Lipid 2 refers to the high *Tm* lipid): *T1L1* (lipid1 tail1 length), *T2L1* (lipid1 tail2 length), ST1 (lipid 1 saturation type), NU1 (lipid 1 number of unsaturated carbons), *MW1* (lipid1 molecular weight), *TM1* (lipid1 melting temperature), *MF1* (lipid1 mole fraction), *T1L2* (lipid2 tail1 length), *T2L2* (lipid2 tail2 length), ST2 (lipid 2 saturation type), NU2 (lipid 2 number of unsaturated carbons), *MW2* (lipid2 molecular weight), *TM2* (lipid2 melting temperature), *MF2* (lipid2 mole fraction), and *T* (temperature of the system). The response variable was the *phase* of the mixture, and it is defined in **Table 7**.

Table 7

Definition of response variable (phase of the mixture).

Phase #	Abbreviation	Definition
1	*Ld*	Liquid-disordered phase
2	*Lo*	Liquid-ordered phase
3	*Lβ*	Solid-ordered phase (gel)
4	*Ld + Lo*	Coexistence of *Ld* and *Lo* phases
5	*Ld + Lβ*	Coexistence of *Ld* and *Lβ* phases
6	*Lo + Lβ*	Coexistence of *Lo,* and *Lβ* phases
7	*Ld + Lo + Lβ*	Coexistence of *Ld*, Lo, and *Lβ* phases
8	*Crystals + Lo*	Cholesterol monohydrate crystals in equilibrium with a cholesterol-saturated *Lo* phase[73,84]

Figure 13

Phase diagrams of different phospholipid/cholesterol mixtures (initial data).
Number codes: 1 = Ld, 2 = Lo, 3 = Lβ, 4 = Ld+Lo, 5 = Ld+Lβ, 6 = Lo+Lβ, 7 =
Ld+Lo+Lβ, and 8 = Crystals+Lo. A) DLPC/DPPC/CHOL at 24oC;84 B)
POPC/DSPC/CHOL at 23oC;80 C) DOPC/DSPC/CHOL at 23oC;80 D)
DOPC/POPC/CHOL at 23oC;80 E) DOPC/DPPC/CHOL at 28oC;87 F)
DOPC/DPPC/CHOL at 22oC;87 G) DOPC/DPPC/CHOL at 18oC;87 and H)
SDPC/BSM/CHOL at 23oC.79 The phase diagram in H was used as the testing set. Lipid
abbreviations: DLPC: 1,2-dilauroyl-sn-glycero-3-phosphocholine, DPPC: 1,2-
dipalmitoyl-sn-glycero-3-phosphocholine, POPC: 1-palmitoyl-2-oleoyl-glycero-3-
phosphocholine, DSPC: 1,2-distearoyl-sn-glycero-3-phosphocholine, DOPC: 1,2-
dioleoyl-sn-glycero-3-phosphocholine, SDPC: 1-stearoyl-2-docosahexaenoyl-sn-glycero-
3-phosphocholine, BSM: Sphingomyelin (Brain).

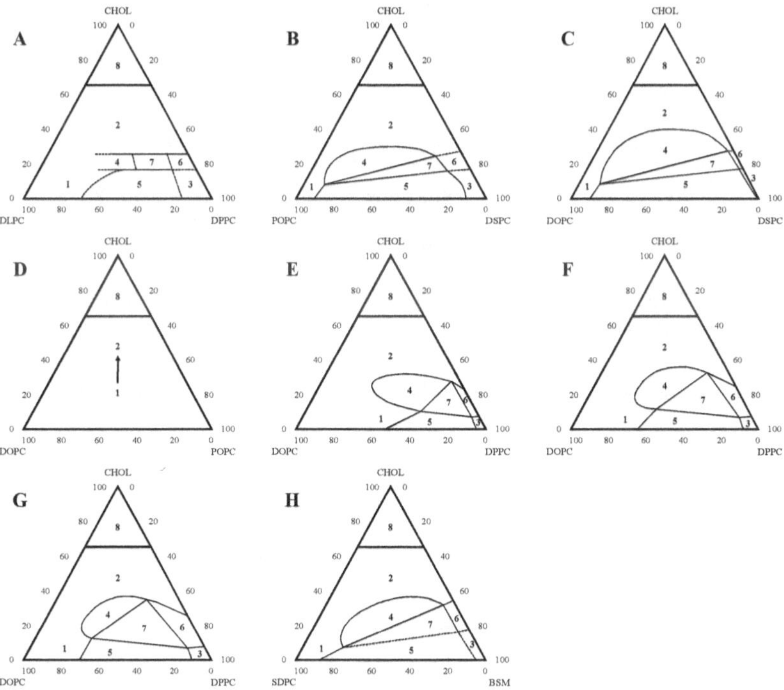

Data Preprocessing. To efficiently utilize machine learning for extracting information from the dataset, a series of preprocessing steps including data balancing and data transformation were performed. Specifically, in the phase diagram dataset, the class labels of the response variable are not evenly distributed (i.e., there are more points in one phase compared to others), which could result in difficulties in the learning algorithms (see **Table 8**). To overcome this unbalancing issue, the well-known synthetic minority over-sampling technique (SMOTE) was used.[113] The general idea behind the SMOTE is to artificially generate new samples of classes from datasets that are not equally represented using their nearest neighbors. This procedure is called over-sampling and is controlled by two parameters of OS and NN. OS controls the size of over-sampling, and NN controls the number of neighbors considered for generating new samples. SMOTE also uses an under-sampling mechanism for the classes with the majority sample. The size of the under-sampling is controlled by the US parameter. Overall, the use of SMOTE will lead to having a more balanced dataset concerning all classes.

Table 8

Percentages of the class labels of the response variable.

Phase	Ld	Lo	$L\beta$	$Ld+Lo$	$Ld+L\beta$	$Lo+L\beta$	$Ld+Lo+L\beta$	$Crystals+Lo$
Percentage %	19.82	36.14	1.72	9.87	11.28	2.44	6.50	12.23

Data Transformation. In the melting temperature dataset, four features BT (backbone type), ChT (acyl chain type), ST (saturation type), and HCh (head charge) were categorical and translated into dummy features accordingly to be implemented in the ANN model. In the phase diagram dataset, *ST1* (lipid 1 saturation type) and *ST2* (lipid 2 saturation type) were categorical which are translated into dummy features and the rest of the features were treated numerically. In both data sets, *Max-Min* normalization was performed to scale the value associated with the features to the range of zero to one.

Design of Elements

Most of the machine learning approaches require the setting of hyper-parameters before the training is initiated. The hyper-parameters determine the structure and size of the machine learning models. Here, an analytical technique was used to fine-tune the hyper-parameters to obtain the best outputs for each technique, as outlined below.

Hyper-parameter Tuning for ANN. The grid search technique was used to tune the hyper-parameters of ANN. This is a technique that builds extensive models based on different combinations of the values of the hyper-parameters and selects the combination that results in the best performance. For this purpose, a multi-layer perceptron neural network model with two or three hidden layers (HLs) was examined. The transfer function for the neurons was set to be identity, logistic sigmoid, or hyperbolic tangent. For the learning process, momentum, conjugant gradient, and Levenberg-Marquardt algorithms were investigated as suggested elsewhere.[114] The number of neurons in each HL was varied from 1 to 9 in increments of 1. After comparing all possible combinations of the hyper-parameter values, the best combination that yielded the minimum root mean square error (RMSE), defined in **Equation 7**, was: two HLs, four neurons in HL1, two neurons in HL2, hyperbolic tangent transfer function, and momentum as the learning algorithm. This combination was selected for building the predictive model.

$$RMSE = \sqrt{\frac{\sum_{i=1}^{n}\left(Tm_{output} - Tm_{desired}\right)^2}{n}}$$

Equation 7

Hyper-parameter Tuning for RF and SVM. The performance of the classification models is highly dependent on the value of their hyper-parameters. Here, the hyper-parameters of RF (*NT*, and *NV*), and SVM (*Cost* and *Gamma*) along with three hyper-parameters of SMOTE (*OS*, *US*, and *NN*) needed to be tuned. Five level values were considered for each hyper-parameter, resulting in 5^5 = 3,125 permutations for a full factorial design. Because of the random nature involved in these algorithms, multiple executions for each experiment were required to achieve robustness in the results. Assuming 20 runs for each experiment, this results in a total of 3,125×20 = 62,500 executions, which is computationally exhaustive. Thus, the Taguchi method was applied to reduce the number of experiments to determine the proper value for each parameter.[115] Given the five factors, each with five levels (**Table 9**), an L_{25} orthogonal array was designed and reported in **Table C3** (Appendix C).

Table 9

The value of hyperparameters for each level in RF and SVM models.

| Level | Hyper-parameter (Factor) | | | | | | | | | |
| | RF | | | | | SVM | | | | |
	OS	US	NN	NV	NT	OS	US	NN	Cost	Gamma
1	100	100	1	3	100	100	100	1	2^{-3}	2^{-9}
2	200	200	5	4	250	200	200	5	2^{-1}	2^{-7}
3	300	300	10	5	500	300	300	10	2^{1}	2^{-5}
4	400	400	15	6	750	400	400	15	2^{3}	2^{-3}
5	500	500	20	7	1000	500	500	20	2^{5}	2^{-1}

The Taguchi method determines an acceptable value for each factor level by maximizing a signal-to-noise (S/N) ratio. Signal represents the mean of the response variable (objective function) and noise denotes standard deviation. The objective function can be in three types: the larger-the-better, the smaller-the-better, and nominal-is-best.[116] In this study, the larger-the-better response variable was applied, as it was desirable to maximize the overall accuracy of the classification models. The S/N ratio is calculated according to **Equation 8**:

$$S/N = -10 \times log_{10}(\frac{1}{e} \times \sum_{i=1}^{e} \frac{1}{ACC^2}) \qquad \text{Equation 8}$$

Here, e is the number of experiments and ACC is the overall accuracy of the classification model. The ACC is calculated by adopting a k-fold cross-validation mechanism. For this purpose, the training dataset is partitioned into k subsets with equal sizes. Then, given the parameters' value corresponding to each experiment, $k\text{-}1$ partitions are considered for training, and one partition is used for validating the trained models. In this study, a 7-fold cross-validation method was utilized. To do so, six out of the seven phase diagrams (**Figure 13A** to **G**) were considered as the training set while one phase diagram was considered as the validating dataset. This process was repeated seven times in such a way that each phase diagram appears one time as the training dataset and one time as the validating dataset. Finally, the averaged ACC was obtained and used in S/N calculation. The calculated S/N ratios for both RF and SVM methods are shown in **Figure C3** (Appendix C). The factor level that yielded the highest S/N ratio for each hyper-parameter is selected as the optimal value. As depicted in **Figure C3** (Appendix

C), the best factor levels for *OS*, *US*, *NN*, *NV*, and *NT* as the hype-parameters of SMOTE-RF are 5, 4, 3, 5, and 5, respectively. **Figure C3** (Appendix C) also illustrates that in the SMOTE-SVM method, the optimal factor levels for *OS*, *US*, *NN*, *Cost*, and *Gamma* are 4, 3, 2, 5, and 4, respectively. The optimal value corresponding to each factor are also reported in **Table 10**.

Table 10

The optimal values of hyperparameters for the RF and SVM algorithms.

RF						SVM				
OS	US	UN	NV	NT		OS	US	NN	Cost	Gamma
500	400	10	7	1000		400	300	5	2^5	2^{-3}

Evaluation Metrics

In the context of classification problems with multi-classes, the overall accuracy of the model can be measured either by the micro-average accuracy or the macro-average accuracy. The macro-average treats all classes equally, as opposed to the micro-average that treats all samples equally. Given the unbalance phase diagram dataset (see **Table 8**), the performance of the phase diagram classification models is better reflected by the micro-average criteria reported in **Equation 9**, **Equation** 10, and **Equation** 11. The *precision* and *recall* were also reported for each class (C_i, where $i \in \{1,2,...,8\}$). These performance metrics are presented below according to the report of Sokolova and colleagues:[117]

$$\text{Recall} \qquad Rec_{C_i} = \frac{TP_{C_i}}{TP_{C_i} + FN_{C_i}} \qquad \text{Equation 9}$$

$$\text{Precision} \qquad Pre_{C_i} = \frac{TP_{C_i}}{TP_{C_i} + FP_{C_i}} \qquad \text{Equation 10}$$

$$\text{Overall Accuracy} \qquad ACC = \frac{\sum_{i=1}^{N} TP_{C_i}}{\sum_{i=1}^{N}(TP_{C_i} + FN_{C_i})} \qquad \text{Equation 11}$$

In the above equations, TP_{C_i} (true positive) indicates the number of correctly classified instances of class C_i. In the case of occurring errors, FN_{C_i} (false negative) is the number of instances that belong to class C_i, but are predicted as other classes, and FP_{C_i} (false positive) is the number of instances do not belong to class C_i but are labeled as C_i.

Results and Discussion

Prediction of Phospholipid Melting Temperature

The ANN algorithm was used to predict lipid melting temperature based on the chemical structure of lipids. Various ANN models were evaluated for this purpose and an optimized model was selected, based on minimized root mean square error (RMSE) and maximized accuracy. This model resulted in an RMSE value of 0.06 and an accuracy of 95.42%. **Figure 14A** shows the performance of the selected ANN model in predicting the melting temperature of 15% of phospholipids randomly selected in the *Tm* dataset (selected lipids are marked with * in **Table C1**, Appendix C). From **Figure 14A**, it can be observed that the predicted values are close to those reported in the literature. This is further confirmed by **Figure 14B**, in which the predicted and reported *Tm* values have been plotted. An R^2 value of 0.96 confirms the linearity of the plot and the accuracy of

the ANN model.

Figure 14

Accuracy of the ANN model to predict phospholipids melting temperature.
A) Tm desired (blue solid line) compared to Tm predicted (red dashed line) for each
tested sample. B) Tm predicted vs. Tm desired for each tested sample. The red line in A
represents the predicted values while the blue line displays the desired values. For
example, for sample seven, the desired value is 0.72 (i.e., Tm = 40.0°C) while the
predicted value is 0.73 (i.e., Tm = 42.2°C). In both figures, the y-axis on the left is the
normalized melting temperature, while on the right is the actual melting temperature. All
Tm values were normalized based on the training dataset at the time of building the
model.

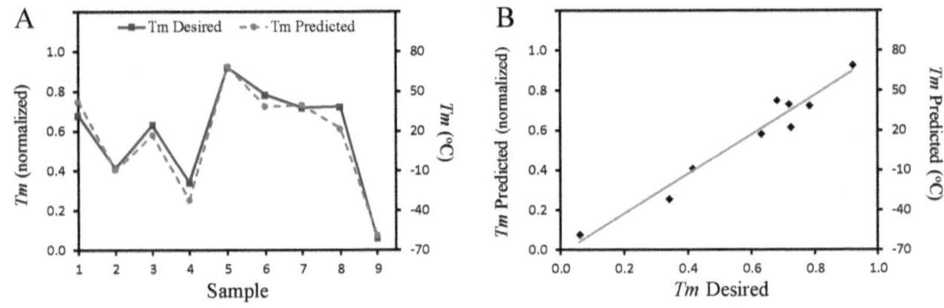

Since the selected model was able to efficiently predict the *Tm* of the tested

samples, it was used to predict the *Tm* values for lipids for which experimental results are

currently not available. To this aim, the lipid headgroup, chain length, and saturation

level were altered, and, in each case, the *Tm* values were generated using the ANN

model. Representative results are presented in **Table 11**, while the complete set of results

can be found in **Table C4** (Appendix C).

Table 11

A representative table of predicted Tm values from the best ANN model.
Predicted values are for lipids not experimentally tested in the literature.

Lipid	Tm (°C)
6:0 PG	-5.4
8:0 PG	-5.3
10:0 PG	-4.3
15:0 PG	35.6
17:0 PG	55.6
14:1 (Δ9-Cis) PC	-46.9
14:1 (Δ9-Trans) PC	25.6
16:1 (Δ9-Trans) PC	31.5
18:0 PA	72.0
18:2 PA	-41.7
02:0 SM (d18:1/2:0)	-5.3
06:0 SM (d18:1/6:0)	-4.9
12:0 SM (d18:1/12:0)	8.9

It can be observed from **Table 11** that while the data generated by the model have not been experimentally confirmed, the trends comply with what is expected from the *Tm* of lipids based on their chemical structure. In all cases, increasing the number of carbons in the acyl chains led to an increase in *Tm*. For example, a *Tm* of -5.3°C was reported for 6:0 phosphatidyl glycerol (PG), which increased consistently as the number of carbons increased, leading to a melting temperature of 55.6°C for 17:1 PG. A similar trend is observed in the case of sphingomyelins. In addition, consistent with the literature, increased unsaturation reduced the melting temperature of lipids.[106,118] For example, while a *Tm* of 72.0°C was reported for 18:0 palmitic acid (PA), the addition of two *cis* double bonds in 18:2 PA reduced the melting temperature to -41.7°C. *Cis* double bonds decreased the *Tm* more than *trans* double bonds as observed in the case of 14:1 (Δ9-*Cis*) phosphatidylcholine (PC) and 14:1 ($\Delta 9$-*trans*) PC, which is also in agreement with the literature.[106,118] Note that the *Tm* for PE lipids has not been predicted using the model given the complex polymorphic phase behavior of these lipids and its dependence on water content.[119,120]

Prediction of Phase Diagrams

RF and SVM algorithms, using the hyperparameters presented in **Table 10**, were used to predict the ternary phase diagrams of phospholipids/cholesterol mixtures. Before predicting the phase diagrams for phospholipid/cholesterol mixtures, for which experimental data was not available, the models were tested for a mixture of SDPC/BSM/CHOL. The phase diagram for this system has previously been reported (**Figure 13H**) and was developed using the RF and SVM models to examine the accuracy of the diagrams predicted by each model.

As can be discerned from **Figure 15**, with more details in **Figure C4** (Appendix C), the RF algorithm generated a better prediction of the phase diagram for the tested lipid mixture compared to the SVM algorithm. This is likely due to the SVM algorithm being less sensitive to non-normal distribution of the data, while the RF model is less sensitive to outliers and multicollinearity among the predictor variables. Importantly, the phase diagram generated by the RF model was in close agreement with the literature results (**Figure 15A** and **B**). While minor differences were observed regarding the size of the phases, the phase boundaries closely mimicked those determined experimentally by Konyakhina and colleagues.[79]

Figure 15

The RF and SVM predicted phase diagrams vs reported experimentally.
*The phase diagram of SDPC/BSM/Chol mixture as **A)** reported by Konyakhina et al.[79]*
*and predicted by **B)** the RF model and **C)** the SVM model. Number codes: 1: Ld, 2: Lo, 3:*
Lβ, 4: Ld+Lo, 5: Ld+Lβ, 6: Lo+Lβ, 7: Ld+Lo+Lβ, and 8: Crystals+Lo.
Lipid abbreviations: SDPC: 1-stearoyl-2-docosahexaenoyl-sn-glycero-3-phosphocholine,
BSM: Sphingomyelin (Brain).

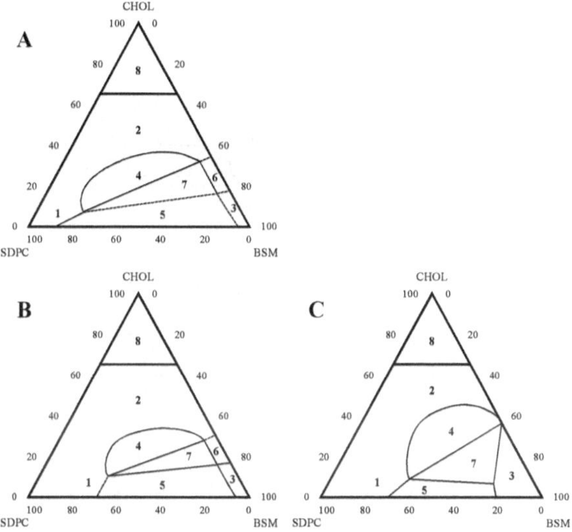

The performance of the models was further assessed using a confusion matrix

(Table 12). The confusion matrix is a tabular representation of the classification results.

The columns of the confusion matrix stand for the actual (desired) phase labels and the

rows stand for the predicted phase labels. The values reported in the main diagonal of the

confusion matrix represent the number of correctly classified mixtures. For example, the

RF model correctly predicted 537 data points to belong to the *Ld* phase, but incorrectly predicted 90 points to be in the *Lo* phase, resulting in 86% accuracy. As can be seen in the confusion matrix, considering *recall* and *precision* as the two-performance metrics, the RF predictive model performs better than the SVM model in almost all classes. The SVM model only showed slightly better performance for the *Ld* phase in terms of *recall* and *precision*. However, the RF was the superior model in terms of *overall accuracy*.

Table 12

The confusion matrix for the RF and SVM predicted phase diagrams.
(1: Ld, 2: Lo, 3: Lβ, 4: Ld+Lo, 5: Ld+Lβ, 6: Lo+Lβ, 7: Ld+Lo+Lβ, and 8: Crystals+Lo)

	Random Forests (RF)								Support Vector Machines (SVM)							
Predicted ↓ / Desired →	1	2	3	4	5	6	7	8	1	2	3	4	5	6	7	8
1	**537**	0	0	158	156	0	36	0	**605**	44	0	203	115	0	38	0
2	90	**1584**	0	63	0	47	19	6	0	**1170**	0	0	0	0	0	0
3	0	0	**85**	0	10	0	0	0	0	0	**105**	0	180	78	19	0
4	0	2	0	**498**	0	0	140	0	22	317	0	**512**	1	0	192	0
5	0	0	16	0	**658**	0	38	0	0	0	0	0	**47**	0	18	0
6	0	0	4	0	0	**58**	4	0	0	0	0	0	0	**0**	0	0
7	0	0	0	0	9	10	**299**	0	0	15	0	4	130	37	**269**	0
8	0	0	0	0	0	0	0	**624**	0	40	0	0	0	0	0	**630**
Recall	0.86	1.00	0.81	0.69	0.79	0.50	0.56	0.99	0.96	0.74	1.00	0.71	0.49	0.00	0.50	1.00
Precision	0.61	0.88	0.89	0.78	0.92	0.88	0.94	1.00	0.60	1.00	0.27	0.49	0.96	NaN	0.59	0.94
Overall Accuracy	**0.84**								**0.72**							

Since the RF algorithm was able to accurately predict the ternary phase diagram for the SDPC/BSM/Chol mixture, it was used to generate phase diagrams for lipid mixtures for which such diagrams have not yet been developed. As a first approach, this algorithm was used to investigate the effect of temperature on (i) the phase behavior of a DOPC/DPPC/CHOL mixture, for which the phase diagrams at a range of temperatures have been reported,[87] and (ii) the phase behavior of a POPC/PSM/CHOL mixture, for which the phase diagrams have been reported,[95] but with less experimental points than the phase diagrams presented in **Figure 13**. The phase diagrams for both systems were predicted at a high (37°C) and low temperature (15°C), allowing for examination of the accuracy of the model, as well as its robustness with respect to temperature. These results are shown in **Figure 16**. Detailed data points for each phase can be found in **Figure C5** (Appendix C).

Changing the temperature resulted in significant differences in the phase diagrams generated using the RF algorithm. When the temperature was reduced (**Figure 16A** and **Figure 16C**), the boundary of the Ld phase (shown by the number 1 in the diagram) was moved to the left. This was accompanied by a significant increase in the $Ld+L\beta$, $Lo+L\beta$, and $Ld+Lo+L\beta$ phases (shown by numbers 5, 6, and 7, respectively). All these changes are expected. This is because lowering the temperature is expected to order the lipids, due to the lowered kinetic energy, as has been reported for other phase diagrams acquired experimentally at various temperatures.[93,95]

Figure 16

The effect of temperature on the RF generated phase diagram.
*The phase diagram for DOPC/DPPC/CHOL system at the temperature of **A)** 15°C, and*
***B)** 37°C. The phase diagram for the POPC/PSM/CHOL system at **C)** 23°C, and **D)** 37°C.*
Number codes: 1: Ld, 2: Lo, 3: Lβ, 4: Ld+Lo, 5: Ld+Lβ, 6: Lo+Lβ, 7: Ld+Lo+Lβ, and
8: Crystals+Lo. Lipid abbreviations: DOPC: 1,2-dioleoyl-sn-glycero-3-phosphocholine,
DPPC: 1,2-dipalmitoyl-sn-glycero-3-phosphocholine, POPC: 1-palmitoyl-2-oleoyl-
glycero-3-phosphocholine, PSM: N-palmitoyl-D-erythro-sphingosylphosphorylcholine.

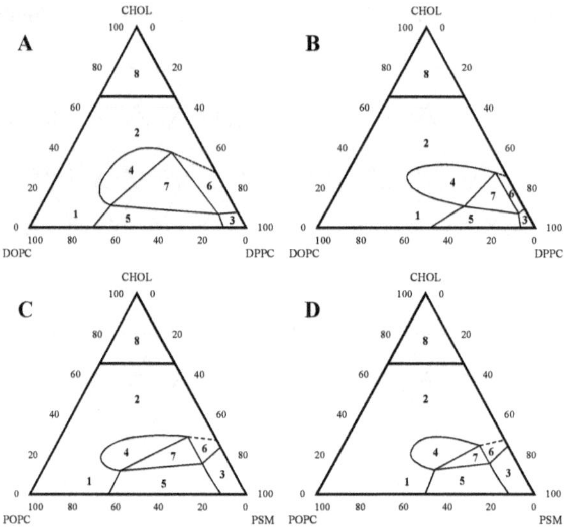

Next, the effect of unsaturation in one vs. both acyl chains was examined. To this aim, the RF algorithm was used to generate a phase diagram for the POPC/DPPC/Chol mixture at 22°C, which was then compared with a DOPC/DPPC/Chol phase diagram from the literature [87] at the same temperature. Here, the presence of a lipid with only one unsaturated acyl chain (POPC) was expected to increase the area of the more ordered phases in the phase diagram, compared to a lipid with two unsaturated acyl chains

(DOPC). This is indeed the trend that is observed in **Figure 17** (**A** and **B**) (detailed data points for each phase can be found in **Figure C6**; Appendix C). The phase diagram for the mixture including POPC showed a substantial increase in the area of the $L\beta$ phase in the composition space, which came at the cost of the $Lo+L\beta$ phase and the three-phase coexistence region, demonstrating an increase in lipid order from the liquid-ordered to the gel phase. Similarly, there was an increase in the size of the $Ld+L\beta$ phase, while the size of the $Ld+Lo+L\beta$ phase was reduced, again indicating an increased gel phase and a decreased liquid-ordered phase. All these effects are expected due to the presence of the more saturated POPC instead of DOPC.

Figure 17

The effect of lipid unsaturation on the RF generated phase diagram.
*A) The phase diagram for the DOPC/DPPC/CHOL mixture from[87] and **B)** the phase diagram for the POPC/DPPC/CHOL mixture as generated by the RF model. The temperature in both diagrams is 22°C. Number codes: 1: Ld, 2: Lo, 3: Lβ, 4: Ld+Lo, 5: Ld+Lβ, 6: Lo+Lβ, 7: Ld+Lo+Lβ, and 8: Crystals+Lo. Lipid abbreviations: DOPC: 1,2-dioleoyl-sn-glycero-3-phosphocholine, POPC: 1-palmitoyl-2-oleoyl-glycero-3-phosphocholine, DPPC: 1,2-dipalmitoyl-sn-glycero-3-phosphocholine.*

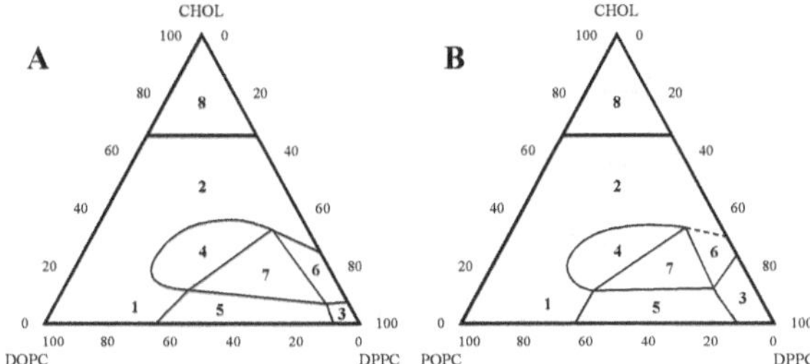

Conclusion

To the best of our knowledge, this is the very first approach in using machine learning for prediction of lipid phase diagrams. While the generated phase diagrams still need to be experimentally tested, the changes in the diagrams generated at different temperatures and different lipid compositions using the RF algorithm suggests that the predicted phase diagrams are robust with respect to changes in the composition or temperature and could potentially be used to generate diagrams for mixtures/conditions that have not yet been experimentally tested. This would be particularly beneficial as developing phase diagrams with accurate tie-lines requires major experimental effort.

This first approach on the use of machine learning for the prediction of lipid phase diagrams could certainly be improved with further optimization. When it comes to optimization, the model can be improved in predicting the $Ld+Lo+L\beta$ phase coexistence region (region 7 in the phase diagrams). This region must be a triangle with 3 straight sides. However, while the model predicts the boundaries for each phase, sometimes the predicted region for this phase does not become an exact triangle. Here, the boundaries for the neighboring regions have been used to identify this region, before plotting the other regions (see e.g., **Figure C4**; Appendix C). However, this approach is somewhat arbitrary and affects the correct prediction of the three-phase diagram. Future efforts should focus on improving the model to ensure that this phase is represented by a triangle. In addition, the lack of a clear boundary in the transition from the Ld to the Lo regions (i.e., region 1 to region 2) creates an issue in the training of the model. Supervised machine learning models require that all classes (in this case phases) are bounded. Thus, an artificial boundary here has been assumed from the plait point of the *Ld+Lo* phase and horizontally to the left side of the triangle (see **Figure C1**; Appendix C). This issue is not solved by better optimization as it is related to the fact that not all phase transitions are first order.

The model can also be improved once further experimental data are available for better training of the algorithm. While the model for *Tm* prediction has been trained with the available *Tm* data, the available experimental data are not balanced. For example, there is more *Tm* data available for lipids with the PC headgroup compared to other headgroups. Also, in unsaturated lipids, there is more data available for lipids with cis unsaturation compared to trans. The availability of such data will help better train the model, leading to more accurate results. The lack of extensive experimental results leads to limitations in the model. For example, while it is possible to use the model to generate phase diagrams at any temperature, the diagrams for temperatures hugely outside the range of 18°C to 28°C would not be highly reliable as the training dataset is only available in this temperature range. This problem could be solved once experimental data at such temperatures are available to train the algorithms. A similar issue exists regarding the phase diagrams for lipids including multiple double bonds in one acyl chain. This is because currently, to the best of our knowledge, no experimental phase diagrams exist for mixtures in which one of the lipids contains multiple double bonds. In other words, and as expected, the proposed machine learning algorithm is only as good as its training dataset, and experiments and machine learning algorithms need to go hand-in-hand.

In conclusion, the current study is a first approach in using machine learning for the prediction of melting temperatures and phase diagrams. The algorithms developed in this study accurately predict melting temperature and phase diagrams for molecules and mixtures for which such data is reported and generate results that are in line with theoretical predictions for systems that are not yet explored. However, the algorithms can be improved by including more features and further training once more experimental results become available. While experiments will undoubtedly continue to hold an invaluable place in biomembrane research, it might be possible to utilize machine learning algorithms for systems for which data is not yet available.

Chapter 5: Modeling and Predicting Performance of Corrosion Inhibitors

The material presented in this chapter is put together as a paper: "Aghaaminiha, M.; Mehrani, R., Colahan, M; Brown, B.; Singer, M.; Nesic S.; Vargas, M., S.; Sharma, S. Machine Learning Modeling of Time-Dependent Corrosion Rates of Carbon Steel in Presence of Corrosion Inhibitors. *Corrosion Science Journal* **2021**". [under review]

Introduction

Surfactant corrosion inhibitors are amphiphilic organic molecules with a polar head group and a hydrophobic tail that are commonly used to mitigate internal corrosion of oil and gas pipelines.[121–123] These molecules are injected in parts-per-million (ppm) concentrations in a continuous or semi-continuous manner. It is understood that corrosion inhibitor molecules adsorb at the metal-water interfaces, with their adsorption chiefly governed by the strong affinity of the polar group for the metal surface as well as hydrophobic interactions between their tails.[121–126] It is important to ascertain the optimal dosage as well as the frequency of the doses of corrosion inhibitors needed to minimize corrosion of the pipelines.[127] For this purpose, extensive experimentation is carried out wherein the corrosion rates of steel specimen are measured in different environmental conditions and by using different dosages and dose-schedules of inhibition. These experiments are time-consuming and costly but are nevertheless necessary, in the absence of reliable theoretical models.

While mechanistic models of corrosion prediction in the absence of inhibitors have been developed and implemented for commercial purposes,[128–131] there has been little success in incorporating the effect of corrosion inhibitors in these models. One difficulty is that the exact mode of action of corrosion inhibitors remains unclear.[122] Secondly, the efficiency of the inhibitors depends on a large number of environmental factors, such as temperature, water chemistry, flow rate, etc.[132] Thirdly, different inhibitor molecules are observed to perform optimally under different conditions for reasons unknown.[133–136] Lastly, corrosion inhibitor formulations used in the field are mixtures of molecules that supposedly work in synergy. Often, the composition of corrosion inhibitor formulations is not disclosed to the operator/researcher as it is proprietary knowledge belonging to the manufacturers, making it all difficult to model their behavior.

Along with experimental methods, molecular modeling has been employed for elucidating the molecular-level adsorption behavior of corrosion inhibitors.[126,137–142] However, molecular modeling can mostly help with understanding of the adsorption behavior of inhibitor molecules and not how they retard corrosion. Therefore, developing theoretical models to explain the performance of corrosion inhibitors remains a challenging task.

In the last decade, machine learning (ML) has been employed in a number of corrosion-related problems, such as for modeling CO_2 corrosion,[143] detecting corrosion from automated image analysis,[144] modeling corrosion defect growth in pipelines,[145] material inspection,[146] modeling corrosion rate in the marine environment,[147] finding corrosion initiation time of embedded steel in reinforced concrete,[148] predicting electrochemical impedance spectra,[149] modeling pipeline aging,[150] and characterizing the spatial distribution of pitting corrosion.[151]

In this work, we have employed different ML algorithms to model experimental data of time-varying corrosion rates of mild steel specimens, when corrosion inhibitors were added to the system in different concentrations and dose-schedules. We demonstrate that the trained ML models are quite accurate in predicting the time-dependent and steady-state corrosion rates of laboratory experiments.

Methodology

Description of the Experimental Data

The experimental data used come from a series of experiments on corrosion inhibition of mild steel in CO_2 aqueous solutions. In the experiments, the inhibited corrosion rate change over time was measured by using Linear Polarization Resistance (LPR). The experiments were conducted independently at four different laboratories, using two different organic corrosion inhibitors (*CI-1* and *CI-2*). The experimental matrix was designed to cover a large parameter space as shown in **Table 13**. Individual experiments lasted between one and seven days and were conducted at CO_2 partial pressures between 0.5 and 12 bar at temperatures between 80°C and 130°C. In some cases, the pH of the solution was controlled at pH 6 while in others that was not the case. Overall, there were 25 different experimental conditions, with many experiments replicated multiple times (26,855 corrosion rate data points in total).

Table 13

Environmental and operational input variables.
These variables are considered to model corrosion rate as a function of time. The last
four features were added to achieve accurate modeling of the experiments.

Description	Range	Unit	Type
Corrosion inhibitor concentration	[0 – 500]	ppm	Numerical
Exposure duration	[0 – 160]	hour	Numerical
CO_2 partial pressure	[0.51 – 12]	bar	Numerical
Temperature	[80 – 130]	°C	Numerical
Corrosion inhibitor type	{*CI-1, CI-2*}	–	Categorical
Wall shear stress	(20, 277)	Pa	Numerical
Brine ionic strength	(0.615, 2.31, 0.51)	mol/L	Numerical
Brine type	{*A, B*}	–	Categorical
pH	{controlled 6, uncontrolled}	–	Categorical
Prior corrosion inhibitor concentration	[0 – 200]	ppm	Numerical
Initial corrosion rate	[0.05, 34.10]	mm/y	Numerical
Pre-concentration	{0, 1}	–	Categorical
Type of test	{sequential dosing, single dose}	–	Categorical

The corrosion inhibitors were added to the solution at different times, in different

dosages, and by using various dosing schedules. Some of the experiments were

conducted with "pre-corrosion", meaning that the steel specimens were allowed to

corrode for some time, before the first dose of the inhibitor was added. In shorter

experiments, the addition of the inhibitor was done as a single dose, while sequentially

increasing dosing was performed in longer experiments. **Figure 18** shows plots of typical

corrosion rate data taken from two experiments: (A) with sequential dosing of the

inhibitor, starting with 1 ppm and incrementally increasing up to 10 ppm, and (B) when a

single dose of inhibitor was added at 10 ppm. In both cases, inhibitor addition was done

after a period of pre-corrosion of about 4.5 hours. The corrosion rate started out high, at

about 7.2 mm/y in both cases, and remained constant during the pre-corrosion period.
From **Figure 18A**, we can observe that already after the addition of 1 ppm of the
inhibitor, the corrosion rate was cut approximately 10-fold, and that every next addition
reduced the corrosion rate further. At the end of the experiment, the corrosion rate was
about 0.04 mm/y obtained with 10 ppm inhibitor, which amounts to an inhibition
efficiency of 99.4%. Comparing the sequential dosing with the single dose inhibitor
injection, **Figure 18** (**A** and **B**), we can observe that a very similar final corrosion rate
was achieved (approximately 0.04 mm/y), suggesting that the dosing schedule was not an
important variable. All other experiments in the dataset were of a similar nature, even if
the details varied.

Figure 18

Two typical types of corrosion experiments.
*Inhibitor added A) sequential dosing, **B**) single dose. Arrows indicate the concentration
in ppm and the time at which the corrosion inhibitor was injected into the system. In both
experiments: inhibitor type = CI-1, p_{CO2} = 0.51 bar, T = 80 °C, and pH = controlled 6.*

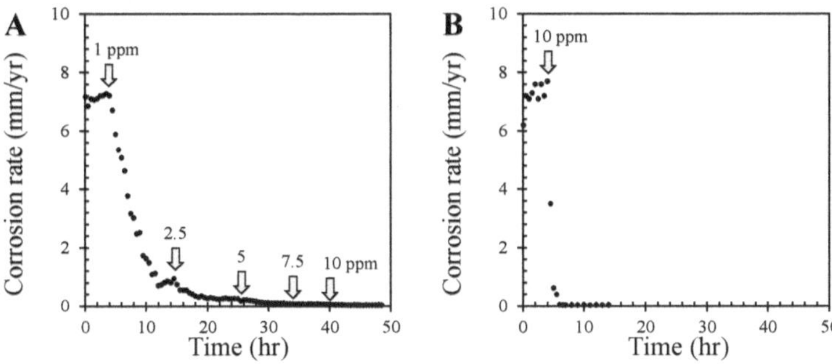

Employed ML Methods

In this work, we have developed several different predictive models of corrosion rates of carbon steel as a function of time, when corrosion inhibitors are added to the system in different dosages and dosing schedules using the different ML algorithms (ANN, RF, SVR, KNN). The ML-based models are useful in predicting corrosion rates at different environmental conditions and alleviate the need for further experimentation. Our analysis shows that RF is the best ML algorithm for this type of modeling for a given data. The details of the implementation are discussed next.

ML Implementation Methodology

To perform prediction using ML algorithms for any data, the following steps need to be taken: (1) data preprocessing, (2) tuning of hyperparameters for several ML models, (3) finding the optimum ML algorithm, (4) training the optimum ML algorithm and testing its performance, and (5) sensitivity analysis to study the response in the outcome as a function of changes in input variables. Each of these steps will be discussed in detail below.

Step 1. Data preprocessing. The first step is to ensure that the dataset does not have any missing and incorrect values.[152,153] Data points with missing values are either removed or the missing values are replaced with the mean or mode or a best guess value of that variable. If a variable has many missing values, then the variable may be removed from the dataset. Different variables have a different range of values and so it is important to rescale the variables. One way is to rescale the variables so that all the values are between 0 and 1. Another way to rescale the variables is to assume that the values are normally distributed and so one can convert them into standard normal variables by subtracting out the mean and dividing by the standard deviation. **Table 13** lists the variables and their ranges in the experimental data of inhibited corrosion rate measurements that we have modeled in this work. The dataset includes eight numerical and five categorical features. Categorical features are converted to dummy variables. For example, if the categorical feature "Brine type" has two possible values, *A* or *B,* then a dummy variable for *A* can be created that takes two possible values {0, 1} with 1 representing brine type *A* and 0 representing brine type *B*. In our work, all numerical variables except *Time* were rescaled to have a standard normal distribution.[153] In the implementation of our ML strategy, we found that for accurate modeling of the experiments, some additional data preprocessing steps were needed. First, the *Time* was set to zero at each injection time of the corrosion inhibitor. Furthermore, the following new input variables were added: (a) prior corrosion inhibitor concentration: this variable indicates the inhibitor concentration present in the system prior to a new dose; (b) initial corrosion rate: this feature specifies the corrosion rate at the beginning of the experiment; (c) pre-concentration: this categorical feature is set to 1 after the first dose of corrosion inhibitor, and is 0 before the first dose; and (d) type of test: this categorical feature is set to 1 if the experiment involved more than one doses of corrosion inhibitor, that is for a sequential dosing experiment, and is set to 0 if there is only a single dose of corrosion inhibitor in the experiment.

Step 2. Tuning hyperparameters. One needs to determine the best hyperparameters for each ML algorithm, that is the hyperparameters that will result in the smallest mean squared error (MSE) during the training.[13] For this step, 75% of the data points were randomly selected and assigned as the training set, and the remaining 25% were hidden from the models and later used as the testing set. The best hyperparameters were found by performing a grid search over a range of hyperparameter values. For each set of hyperparameter values, a five-fold cross-validation method was used on the training set. In 5-fold cross-validation, the training data is divided into 5 subsets. The ML algorithm is then trained using the data in four subsets, and the performance of the trained algorithm is tested on the 5th subset. This procedure is repeated 5 times, each time choosing a different subset for testing the performance.[3,4] Then, the average performance of the ML algorithm (in our case MSE score) on the training set is determined. Using this methodology, the best hyperparameter values of the different ML models (ANN, SVM, RF, and KNN) were determined.

Step 3. Finding the optimum ML model. After tuning the hyperparameters, the performance of the models on the testing set was compared and the model with the best performance, that is the one with the lowest MSE, was selected as the optimum model.

Steps 4. Training the optimum ML model and testing its performance. One useful and practical application of our ML modeling was to be able to predict results of complete experiments of corrosion rates as a function of time for given environmental conditions and dosage schedule of corrosion inhibitors. To check the performance of our optimum ML model, a different strategy was adopted compared to the one used in **Step 3**, where 75% of data points used for training were selected from all the experiments, meaning that none of the individual experiments were completely hidden from the model.

To make it more challenging for the model, the process of training was repeated but this time by using data from randomly selected 21 out of 25 experiments, which formed the training set; the remaining 4 experiments were completely hidden from the model and were later used as the testing set. In cases where an experiment was repeated multiple times and was randomly selected for the testing set, all its replicas were hidden from the model during training.

Step 5. Sensitivity analysis. A trained ML model can also be used to study the behavior of the system as a function of changes in the values of the input features. In our study, the time-varying corrosion rate was predicted as a function of inhibitor type, inhibitor concentration, CO_2 partial pressure, temperature, wall shear stress, and brine type using the trained ML model.

Results and Discussion

Identifying the Best ML Model

The ranges of the hyperparameters that were tested for each of the ML algorithms in this study are listed in **Table 14**. For ANN, the number of hidden layers (*nHLs*) and the number of nodes per hidden layer (*nNodes*) were the two hyperparameters that were varied. In SVR, two hyperparameters were varied: the kernel coefficient, γ, and the regularization parameter, *cost*. As discussed above, in SVR, the input variables are transformed to higher dimensions using kernel functions. We have used Gaussian functions as the kernels. Here, γ refers to the inverse of the standard deviation or the spread of the Gaussian functions used. The *cost* parameter controls how many kernel functions are employed in the SVR. For RF, we choose the number of decision trees, *nTR*, and the maximum fraction of input variables that can be used to create a decision tree, *mFF*, as the two hyperparameters. For KNN, the value of K, that is, the number of nearest neighbors that one should consider is taken as the hyperparameter. Alongside, we studied the performance of KNN models when equal weight is given to every neighboring data point (*weight* set to "uniform") and when nearest neighbors are given more weight than the farther ones (*weight* set to "distance"). It should be noted that for each ML algorithm, only the hyperparameters that are understood to have the largest effect on the performance were studied, while choosing the default suggested values of the other hyperparameters.

Table 14

Ranges of values for the employed ML algorithms' hyperparameters.
(nHLs: number of hidden layers, nNodes: number of nodes per hidden layer; γ-gamma:
kernel coefficient, cost: regularization parameter; nTR: number of trees, mFF: maximum
fraction of features to be split for a decision tree; and K: number of neighbors)

ANN		SVM		RF		KNN	
nHLs	*nNodes*	*γ*	*cost*	*nTR*	*mFF*	*K*	*weights*
1	2	1.0	1	10	0.6	3	uniform
2	4	0.1	5	50	0.7	4	distance
3	6	0.01	10	100	0.8	5	
4	8	0.001	100	200	0.9	6	
5	10	0.0001	1000	500	1.0	7	

The results of the 5-fold cross-validation of the ML algorithms for different values of the hyperparameters are shown in **Figure 19**.

Figure 19

Studied ML methods' hyperparameters tuning strategy.
Performance of A) Artificial Neural Network (ANN), B) Support Vector Machines (SVM),
C) Random Forest (RF), and D) K Nearest Neighbors (KNN) on experimental data of
corrosion rate for different values of the hyperparameters. Each point is the average of
20 independent iterations. Lines are guides for readability.

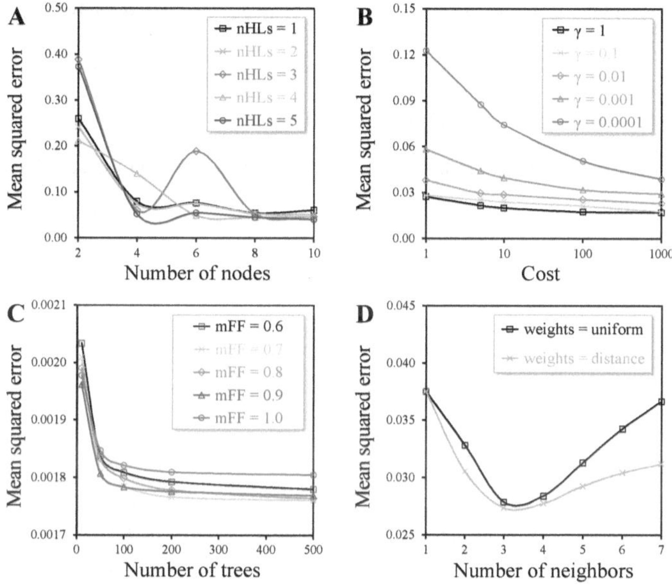

For ANN (**Figure 19A**), it was observed that the mean squared error (MSE)

decreased with the increase in the number of nodes per hidden layer. The decrease in the

MSE became gradual beyond 6 nodes. Increasing the number of nodes beyond 8 resulted

in a slight increase in the MSE. Therefore, 8 nodes per hidden layer with 4 hidden layers

was the optimum architecture. In the case of SVR (**Figure 19B**), it was observed that the

performance was sub-optimal for small values of γ (< 0.001) and was strongly dependent on the *cost*. For $\gamma > 0.01$, the performance showed significant improvement and was not overly sensitive to the *cost*. The MSE was observed to decrease with the *cost*, and so the SVM model with *cost* and γ set to 1000 and 1, respectively, was the optimum SVR model. In the case of RF model (**Figure 19C**), the best performance (lowest MSE) was observed when the *mFF* was set to 0.7. Beyond 200 decision trees, some improvement in the performance was observed and so the RF model with 500 trees was the optimum RF model. In the case of KNN (**Figure 19D**), a V-shaped curve of the MSE was obtained, implying the $K = 3$ as the optimum choice. Also, having the *weight* based on the distance rather than having a uniform weight for all neighbors yielded lower MSE. The optimum set of hyperparameters for each ML model are reported in **Table 15**.

Table 15

Optimum values of the hyperparameters.
Note that the reported MSEs are the average of 20 independent iterations. (nHLs:
number of hidden layers, nNodes: number of nodes per hidden layer; γ-gamma: kernel
coefficient, cost: regularization parameter; nTR: number of trees, mFF: maximum
fraction of features to be split for a decision tree; and K: number of neighbors)

	ANN		SVM		RF		KNN	
	nHLs	*nNodes*	γ	*cost*	*nTR*	*mFF*	*K*	*weights*
Hyperparameters	4	8	1	1000	500	0.7	3	distance
MSE	0.048±0.010		0.017±0.001		0.002±0.001		0.027±0.001	

It was found that RF is the best algorithm among those studied with the MSE of 0.002 ± 0.001 (**Figure 20A**). Therefore, RF was selected as the ML model for our investigations of corrosion rate as a function of time. The parity plot of the predictions of the RF model of the corrosion rate on the testing set, which is randomly selected as 25%

of the whole 26,855 data points, is shown in **Figure 20B**. The parity plot shows that the

RF model is quite accurate in predicting the experimental results of corrosion rate

measurements.

Figure 20

Comparison of studied ML model performance on the testing set.
A) Comparison of performance of the best model for each ML model. Error bars are
obtained by performing 20 independent iterations of each model. B) Parity plot between
the predicted corrosion rates (mm/y) by the RF model and the experimental values; The
reported MSE value is the average of 20 independent iterations of the RF model. In both
figures, predictions are made on the testing set which is randomly selected as 25% of the
available 26,855 data points.

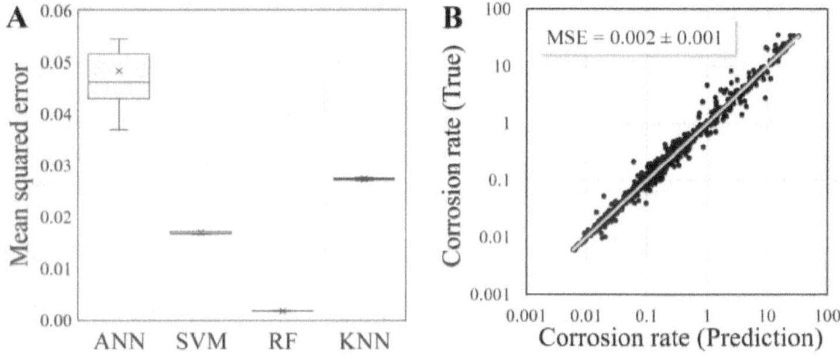

Predictions of Time-dependent Corrosion Rates Using RF

The first set of results presented here are from experiments with sequential dosing

of the inhibitor. **Figure 21** shows eight experiments done with pre-corrosion and their

prediction by the RF model. Corrosion rates are plotted on a logarithmic scale, to be able

to show on the same graph, both the high corrosion rates at the beginning of the

experiment before inhibitor injection and the low corrosion rates after inhibition. It can be observed that the initial corrosion rates reflect the severity of the aqueous environment. For example, the conditions of experiments shown in **Figure 21 (A, D, and E)** were more aggressive, having the initial corrosion rates in the range 20 – 30 mm/y, which can be attributed primarily to the higher partial pressures of CO_2 in these experiments (p_{CO_2} = 2.5, 1.57, and 2.5 bar respectively), This is clear when comparing to the conditions presented in **Figure 21 (B, C, F, G, and H)** where the p_{CO_2} = 0.51 bar and the initial corrosion rates were about 10 mm/y. One would imagine that the final corrosion rates (after inhibition) are related to the inhibitor type and the inhibitor dosing, but this actually is not the case.[1] For example, for both *CI-1*, shown in **Figure 21 (A and B)** and *CI-2*, shown in **Figure 21 (C and D)** the final corrosion rate seems to be more related to the severity of the environment, rather than to the inhibitor type. Similarly, there seems to be no correlation with the inhibitor concentration, as with both high and low final doses, we see both bad and good inhibition, which seems to be related to the severity of the environment and not to inhibitor concentration. For example: in experiments shown in **Figure 21 (A, D, and E)** the final corrosion rate stayed relatively high, and in the experiments shown in the **Figure 21 (D and E)** cases – it never decreased below 0.1 mm/y. This is mostly related to the high p_{CO_2} in those experiments and a higher temperature (106°C vs. 80°C); it is well known that organic corrosion inhibitor performance deteriorates at a higher temperature. In the same experimental condition, shown in **Figure 21D**, we see a substantial scatter of the inhibited corrosion rate results

---

[1] This conclusion is limited to the two inhibitors considered here and is not a general rule.

(within a factor 5 in the 9 "replicated" experiments), which is primarily related to the uncontrolled pH in those experiments and compounded by the severe conditions for inhibition (high p_{CO_2} and high temperature). Finally, there seems to be no influence of brine type or ionic strength on the corrosion rate under any condition used in this study.

When it comes to predictions, we can observe that the RF model was quite accurate in reproducing the experimentally observed values in all cases. This holds true even for the experiments shown in **Figure 21D**, where the large scatter in the experimental results is seen, yet the RF model falls within the scatter band. In that condition, the prediction from the RF model shows some "jumps" in the corrosion rate at times when the corrosion inhibitors were injected into the system, which is clearly an unrealistic behavior of the model, but, nevertheless, the overall prediction from the model remains reasonable. An outlier to some extent is the case presented in **Figure 21E** where the conditions were rather extreme (p_{CO_2} = 2.5 bar wall shear stress of 277 Pa), hence the RF model had more trouble in predicting the inhibited corrosion rate evolution, even if it got the final corrosion rate right.

Figure 21

Sequential dosing experiments prediction vs. experiments.
Corrosion rate as a function of time. RF prediction (red), Experimental data (grey).

	A	B	C	D	E	F	G	H
CI type	*CI-2*	*CI-2*	*CI-1*	*CI-1*	*CI-1*	*CI-2*	*CI-1*	*CI-2*
Exposure dur.	~ 90 hrs.	~ 50 hrs.	~ 50 hrs.	~ 70 hrs.	~ 80 hrs.	~ 80 hrs.	~ 40 hrs.	~ 50 hrs.
p_{CO2}	2.5 bar	0.51 bar	0.51 bar	1.57 bar	2.5 bar	0.51 bar	0.51 bar	0.51 bar
Temperature	80 °C	80 °C	80 °C	106 °C	80 °C	80 °C	80 °C	80 °C
pH	controlled	controlled	controlled	uncontrolled	controlled	controlled	controlled	controlled
Wall shear stress	277 Pa	20 Pa	20 Pa	20 Pa	277 Pa	20 Pa	20 Pa	20 Pa
Brine ion. Str.	0.51 M	0.615 M	2.31 M	2.31 M	0.51 M	0.51 M	0.615 M	2.31 M
Brine type	*A*	*A*	*B*	*B*	*A*	*A*	*A*	*B*

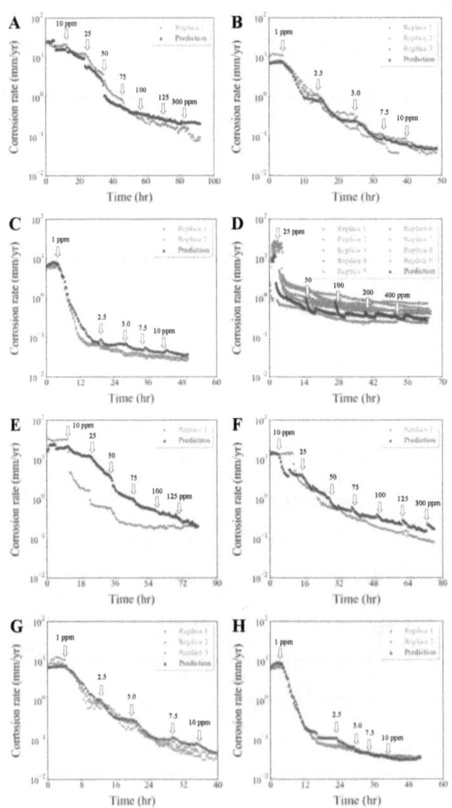

Next, we will look at single-dose experiments with pre-corrosion shown in **Figure 22**. In all cases, we have similar initial corrosion rates, which are quite high: 5 – 10 mm/y, indicating a severe corrosion environment. After inhibition, similar conclusions can be drawn as we had for the experiments with sequential dosing of the inhibitor, discussed above. One can clearly see that neither does the inhibitor type nor its dose directly correlates to the final corrosion rate. The same is true for brine type and ionic strength. The final corrosion rates were low (below 0.2 mm/y) in all cases, except in the experiment shown in **Figure 22C**, which can be explained by the high temperature (132°C), which made it hard for the organic inhibitor to perform well, even if the final dose was extremely high: 100 ppm. Just like in the previous example where sequential dosing was used, the large scatter in the inhibited corrosion rate was obtained in experiments where the pH was not controlled, shown in **Figure 22A**. Despite all this complexity, and scatter, we can argue that the RF model performed quite well, in all cases staying within the error margins of the experimental results. The same conclusion can be reached by inspecting the additional results shown in **Figure 23** obtained without pre-corrosion.

Figure 22

Single dose experiments with pre-corrosion prediction vs. experiments.
Corrosion rate as a function of time. RF prediction (red), Experimental data (grey).

	A	B	C	D	E	F
CI type	*CI-2*	*CI-1*	*CI-1*	*CI-2*	*CI-2*	*CI-2*
Exposure duration	~ 30 hrs.	~ 20 hrs.	~ 22 hrs.	~ 25 hrs.	~ 21 hrs.	~ 28 hrs.
pco2	0.51 bar	0.51 bar	12 bar	0.51 bar	0.51 bar	0.51 bar
Temperature	80 °C	80 °C	132 °C	80 °C	80 °C	120 °C
pH	uncontrolled	controlled	uncontrolled	controlled	controlled	uncontrolled
Wall shear stress	20 Pa	20 Pa	20 Pa	20 Pa	20 Pa	20 Pa
Brine ionic strength	0.615 M	2.31 M	0.615 M	0.615 M	2.31 M	0.615 M
Brine type	*A*	*A*	*A*	*A*	*A*	*A*

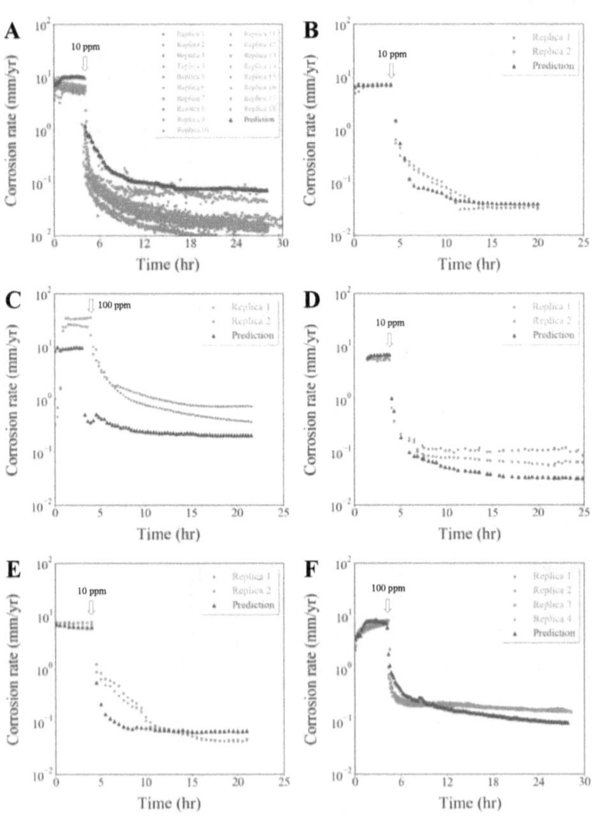

Figure 23

Single dose experiments without pre-corrosion prediction vs. experiments.
Corrosion rate as a function of time. RF prediction (red), Experimental data (grey).

Conditions / Figure	A	B	C	D
CI type	*CI-1*	*CI-1*	*CI-1*	*CI-1*
Exposure duration	~ 145 hrs.	~ 65 hrs.	~ 68 hrs.	~ 68 hrs.
p_{CO_2}	5.0 bar	6.9 bar	6.9 bar	1.57 bar
Temperature	94 °C	130 °C	130 °C	106 °C
pH	uncontrolled	uncontrolled	uncontrolled	uncontrolled
Wall shear stress	20 Pa	20 Pa	20 Pa	20 Pa
Brine ionic strength	0.615 M	0.615 M	0.615 M	2.31 M
Brine type	*A*	*A*	*A*	*B*

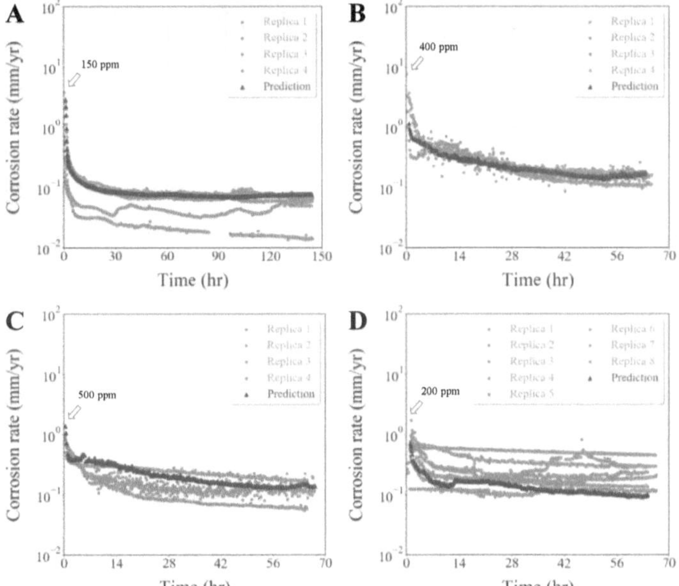

Sensitivity to Input Variables

Trained ML models can predict outcomes for conditions that have not been studied experimentally and thus can be employed to perform parametric studies to investigate how systematically changing one input affects the final prediction. In **Figure 24**, such a parametric study has been performed to determine the effect of changing the corrosion inhibitor type (A) and concentration (B), CO_2 partial pressure (C), temperature (D), wall shear stress (E), and brine type (F), on the corrosion rates. The baseline corrosion scenario was quite severe: high temperature (130°C) and high p_{CO2} = 12 bar. In the simulation, a single dose of inhibitor was added after 4 h of pre-corrosion and the whole simulation lasted 22 h.

Clearly, the type of inhibitor injected, did not seem to matter, as shown in **Figure 24A**, which is consistent with the experimental results shown in **Figure 21**, **Figure 22**, and, **Figure 23**. The effect of inhibitor concentration is shown in **Figure 24B** and is logical – an increase in inhibitor concentration between 100 and 300 ppm leads to somewhat lower inhibited corrosion rates, whereas increasing the dose to 1000 ppm offers little or no additional benefits, according to the RF model. This seems to agree with our understanding that at some high inhibitor bulk concentration, a surface saturation concentration is reached.[154] Overall, it shows that the interpolation capability of the RF model (between 100 and 300 ppm) is excellent, and that this carries over to extrapolation when we look at concentrations up to 1000 ppm, which were not used in the experiments. On the other hand, the extrapolation of results towards lower concentrations (using 10 ppm which was never tested in single-dose experiments with pre-corrosion) gave erroneous results – a better inhibitor performance. This points to the danger of extrapolation using ML models such as RF, when the outcome is unpredictable. One possible remedy for this problem is to use active learning, which is a semi-supervised learning algorithm that continuously updates the training and testing sets and increases the extrapolation capability of ML models.

Similar results are seen when the p_{CO2} was varied, as shown in **Figure 24C**, where the interpolation in the range $p_{CO2} = 0.5$ bar to 12 bar follows the logical trends as seen in the experimental results – higher p_{CO2} leading to higher corrosion rates. However, when the p_{CO2} is increased further (leading to extrapolation) the sensitivity of the corrosion rate to p_{CO2} vanished, what is not an expected result. When it comes to temperature effects, shown in **Figure 24D,** the interpolation capability of the RF model was fine, and the extrapolation to lower temperatures (e.g., 30°C) worked as well, leading to even lower corrosion rates and better inhibition, as expected. Finally, as shown in **Figure 24E** and **Figure 24F**, there was no effect of wall shear stress and brine type, as was deduced from the experimental results.

Figure 24

A parametric study, using a trained RF model, to determine how the time-dependent corrosion rates are expected to vary with some important input variables.
A) inhibitor type, B) inhibitor concentration, C) CO2 partial pressure, D) temperature, E) wall shear stress, and F) brine type. Common condition in all figures, unless otherwise specified is as follow: inhibitor type = CI-1, inhibitor concentration = [0, 100] ppm, exposure duration ~ 22 hrs., p_{CO2} = 12 bar, T = 130 °C, pH = uncontrolled, wall shear stress = 20 Pa, brine ionic strength = 0.615 M, brine type = A.

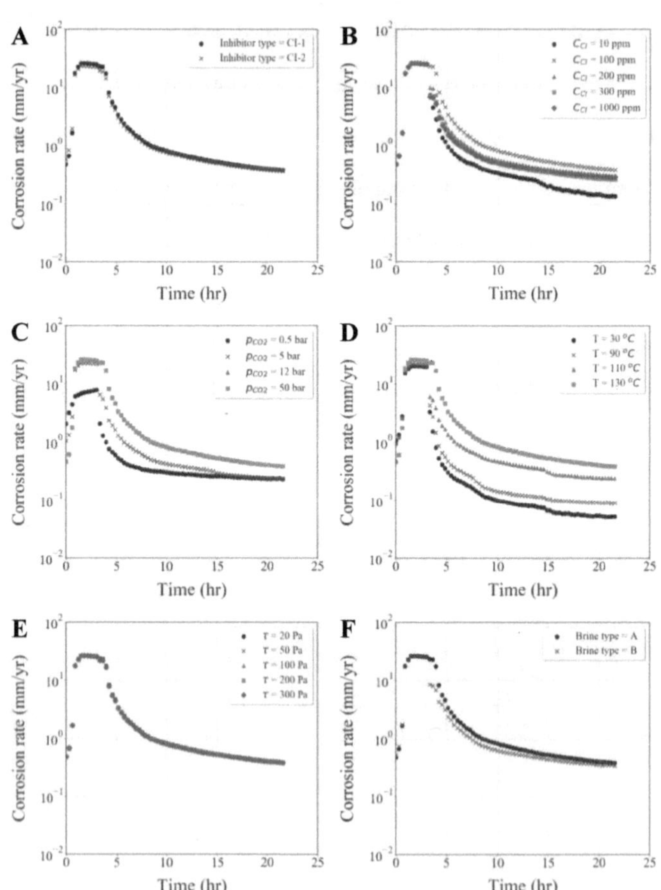

Finally, it is important to mention that throughout the course of this study, the ML modeling effort was continuously discussed with the experts in corrosion science and those with field experience, to ensure that (1) the models were meeting the desired level of accuracy in the predictions and (2) the modeling approach is logical and meaningful when it comes to practical applications. Their input into building, training, evaluating and interpreting results from ML modeling cannot be overemphasized. Therefore, a hypothetical scenario where the role of corrosion experts can be readily replaced by various ML algorithms seems to be both unrealistic and unreliable. On the other hand, ML algorithm are another excellent addition to the toolbox of corrosion experts that can make them more effective, complement other methods they use to reach conclusions and make predictions, and can improve the corrosion experts' overall productivity and quality of performance.

Conclusions

Given that, in an experimental system, there are many environmental variables that affect the performance, an exhaustive study of all possible combinations of the variables is prohibitive. Therefore, a well-trained ML model can be employed to alleviate the need for such experiments.

In this research, the performance of different supervised machine learning models has been compared. Artificial neural networks (ANN), random forest (RF), support vector machine (SVM), and K nearest neighbors (KNN) were tested to predict time-dependent mild steel corrosion rates for different dosages and dosing schedules of corrosion inhibitors. Results show that the RF model outperformed the other models for this dataset. The entire time trend of the corrosion rate of mild steel is quite well predicted by the trained RF model. A trained RF model was also shown to be quite useful in observing how results are expected to vary as the environmental variables are altered.

Chapter 6: Conclusions

In this Ph.D. research, molecular dynamics (MD) and machine learning (ML) techniques are employed to study behavior of biological and metallic interfaces in water. Specifically, the MD simulations work focused on studying the behavior of lipid bilayers. We employed ML methods to model phase behavior of three-components lipid mixtures. Finally, we employed ML methods to model corrosion-related problems. It is noteworthy to mention that the project on the implementation of ML models for corrosion-related problems was sponsored by BP production company, and so proprietary information has not been disclosed. The main conclusions of the research are discussed below.

Research Project 1. *Study the Relationship Between Distribution of Cholesterol and Ordering of Lipids in Lipid bilayers [Discussed in chapter 2].* The plasma membrane of eukaryotic cells is known to be compositionally asymmetric. Certain phospholipids, such as sphingomyelin and phosphatidylcholine species, are predominantly localized in the outer leaflet, while phosphatidylethanolamine and phosphatidylserine species primarily reside in the inner leaflet. While phospholipid asymmetry between the membrane leaflets is well established, there is no consensus about cholesterol distribution between the two leaflets. We have performed a systematic study, via molecular simulations, of how the spatial distribution of cholesterol molecules in different "asymmetric" lipid bilayers are affected by the lipids' backbone, head-type, unsaturation, and chain-length by considering an asymmetric bilayer mimicking the plasma membrane lipids of red blood cells, as well as seventeen other asymmetric bilayers comprising of different lipid types. Our results reveal that the distribution of cholesterol in the leaflets is solely a function of the extent of ordering of the lipids within the leaflets. The ratio of the distribution of cholesterol matches the ratio of lipid order in the two leaflets, thus providing a quantitative relationship between the two. These results are understood by the observation that asymmetric bilayers with equimolar amount of lipids in the two leaflets develop tensile and compressive stresses due to differences in the extent of lipid order. These stresses are alleviated by the transfer of cholesterol from the leaflet in compressive stress to the one in tensile stress. These findings are important in understanding the biology of the cell membrane, especially regarding the composition of the membrane leaflets.

Research Project 2. *Study the Relationship Between Distribution of Cholesterol and Phase Behavior of Lipid bilayers [Discussed in chapter 3].* The Dry and Wet Martini force field in estimation of structural properties of different lipid bilayers representing cells' membrane are fully compared. Then, the behavior of different lipid bilayers at wide range of temperatures and mole-ratios of cholesterol are investigated to successfully access the location and orientation of cholesterol at temperatures higher and lower than the phase transition temperature of bilayers. First, we observed that the Dry Martini force field where water is implicitly considered in the systems is as accurate as Wet Martini force field with explicit water in simulating the lipid bilayers behavior under different conditions. Second, we found that the spatial distribution and orientation of cholesterol is strongly correlated with the phase of the bilayers. In the *ordered* domains, cholesterol molecules are mainly present in the leaflets and orient themselves more parallel to the bilayer normal with the angle less than 30° or greater than 150°. In the disordered domains, cholesterol molecules are predominantly present near the center of the bilayer at the midplane region with the angle between 30° and 150°. In the disordered domains, cholesterol angle with bilayer normal changes linearly with respect to the distance from the center of the bilayer while in the ordered domains this linear relationship is not fully developed. In agreement with the previous experimental studies, we observed that increasing the cholesterol concentration favors ordered domains of the bilayers.

Research Project 3. *Estimation of Phase Diagrams for Three-Component Lipid Mixtures [Discussed in chapter 4].* The plasma membrane of eukaryotic cells is commonly known to contain different lipid domains. The interest in understanding the origin of such domains has led to extensive studies on the phase behavior of mixed lipid systems. Three-component phase diagrams, composed of a high melting temperature (*Tm*) lipid, cholesterol, and a low *Tm* lipid have been valuable in studying lipid phase behavior. However, developing phase diagrams over the entire composition space and with precise tie-lines requires significant experimental effort. In this study, a machine learning approach was used to predict the Tm of lipids and generate phase diagrams from lipid mixtures. First, artificial neural network (ANN) was used for the prediction of *Tm*. The network was trained using available *Tm* data and was able to generate *Tm* values that closely matched literature results for its testing dataset. This model was then used to predict the *Tm* for lipids that have not yet been experimentally tested. Then, random forests (RF) and support vector-machines classifier (SVC) were trained and tested for their ability to predict a test three-component phase diagram. The model from the RF algorithm was able to generate a diagram that closely matched published results. This model was then used to generate phase diagrams for lipid mixtures at various temperatures and various degrees of unsaturation. This machine learning approach to the generation of lipid phase diagrams has the potential to save significant time and resources in studies of lipid phase behavior.

Research Project 4. *Modeling and Predicting Performance of Corrosion Inhibitors [Discussed in chapter 5].* Corrosion inhibitors are injected continuously or semi-continuously in oil and gas pipelines to inhibit internal corrosion. Laboratory experiments are performed for inhibitor selection and to optimize the dosage and dose-schedules of corrosion inhibitors. Although these experiments are time-consuming and costly, they are necessary as it is still very difficult to predict the effect of corrosion inhibitors on the rates of electrochemical reactions in a mechanistic way. However, other approaches can be considered to help anticipate the effect of inhibition, optimize the experimental test matrix, and eventually reduce the number of experiments required, and associated time and cost. In this work, we have employed different supervised machine learning methods, namely, artificial neural network (ANN), random forest (RF), support vector-machines regressor (SVR), and k nearest neighbors (KNN), to model laboratory measurements of corrosion rates of carbon steel as a function of time when corrosion inhibitors are added to the system in different dosages and dose-schedules. We find that RF was the best machine learning algorithm for this modeling. Using the optimized RF model, we show that the entire time profiles of experimentally measured corrosion rates can be predicted quite accurately.

References

(1) Allen, M. P.; Tildesley, D. J. *Computer Simulation of Liquids: Second Edition*; Oxford University Press.

(2) Marrink, S. J.; de Vries, A. H.; Mark, A. E. Coarse Grained Model for Semiquantitative Lipid Simulations. *J. Phys. Chem. B* **2004**, *108* (2), 750–760. https://doi.org/10.1021/jp036508g.

(3) Mitchell, T. M. Does Machine Learning Really Work? *AIMag* **1997**, *18* (3), 11–11. https://doi.org/10.1609/aimag.v18i3.1303.

(4) Géron, A. *Hands-On Machine Learning with Scikit-Learn, Keras, and TensorFlow: Concepts, Tools, and Techniques to Build Intelligent Systems*; O'Reilly Media, Inc., 2019.

(5) Breiman, L. Random Forests. *Machine Learning* **2001**, *45* (1), 5–32. https://doi.org/10.1023/A:1010933404324.

(6) Breiman, L. Arcing Classifier (with Discussion and a Rejoinder by the Author). *Ann. Statist.* **1998**, *26* (3), 801–849. https://doi.org/10.1214/aos/1024691079.

(7) Hornik, K. Approximation Capabilities of Multilayer Feedforward Networks. *Neural Networks* **1991**, *4* (2), 251–257. https://doi.org/10.1016/0893-6080(91)90009-T.

(8) Benitez, J. M.; Castro, J. L.; Requena, I. Are Artificial Neural Networks Black Boxes? *IEEE Transactions on Neural Networks* **1997**, *8* (5), 1156–1164. https://doi.org/10.1109/72.623216.

(9) Bala, R.; Kumar, D. D. Classification Using ANN: A Review. *International Journal of Computational Intelligence Research* **2017**, *13*, 1811–1820.

(10) Gunn, S. R. Support Vector Machines for Classification and Regression. *ISIS technical report* **1998**, *14* (1), 5–16.

(11) Cover, T.; Hart, P. Nearest Neighbor Pattern Classification. *IEEE Transactions on Information Theory* **1967**, *13* (1), 21–27. https://doi.org/10.1109/TIT.1967.1053964.

(12) Xia, S.; Xiong, Z.; Luo, Y.; Dong, L.; Zhang, G. Location Difference of Multiple Distances-Based k-Nearest Neighbors Algorithm. *Knowledge-Based Systems* **2015**, *90*, 99–110. https://doi.org/10.1016/j.knosys.2015.09.028.

(13) Claesen, M.; De Moor, B. Hyperparameter Search in Machine Learning. *arXiv:1502.02127 [cs, stat]* **2015**.

(14) Aghaaminiha, M.; Farnoud, A. M.; Sharma, S. Quantitative Relationship between Cholesterol Distribution and Ordering of Lipids in Asymmetric Lipid Bilayers. *Soft Matter* **2021**, *17* (10), 2742–2752. https://doi.org/10.1039/D0SM01709D.

(15) van Meer, G.; Voelker, D. R.; Feigenson, G. W. Membrane Lipids: Where They Are and How They Behave. *Nat Rev Mol Cell Biol* **2008**, *9* (2), 112–124. https://doi.org/10.1038/nrm2330.

(16) Ingólfsson, H. I.; Carpenter, T. S.; Bhatia, H.; Bremer, P.-T.; Marrink, S. J.; Lightstone, F. C. Computational Lipidomics of the Neuronal Plasma Membrane. *Biophysical Journal* **2017**, *113* (10), 2271–2280. https://doi.org/10.1016/j.bpj.2017.10.017.

(17) Ingólfsson, H. I.; Melo, M. N.; van Eerden, F. J.; Arnarez, C.; Lopez, C. A.; Wassenaar, T. A.; Periole, X.; de Vries, A. H.; Tieleman, D. P.; Marrink, S. J. Lipid Organization of the Plasma Membrane. *J. Am. Chem. Soc.* **2014**, *136* (41), 14554–14559. https://doi.org/10.1021/ja507832e.

(18) Marquardt, D.; Geier, B.; Pabst, G. Asymmetric Lipid Membranes: Towards More Realistic Model Systems. *Membranes* **2015**, *5* (2), 180–196. https://doi.org/10.3390/membranes5020180.

(19) Lorent, J. H.; Levental, K. R.; Ganesan, L.; Rivera-Longsworth, G.; Sezgin, E.; Doktorova, M.; Lyman, E.; Levental, I. Plasma Membranes Are Asymmetric in Lipid Unsaturation, Packing and Protein Shape. *Nature Chemical Biology* **2020**, 1–9. https://doi.org/10.1038/s41589-020-0529-6.

(20) Gibson Wood, W.; Igbavboa, U.; Müller, W. E.; Eckert, G. P. Cholesterol Asymmetry in Synaptic Plasma Membranes: Brain Membrane Cholesterol Asymmetry. *Journal of Neurochemistry* **2011**, *116* (5), 684–689. https://doi.org/10.1111/j.1471-4159.2010.07017.x.

(21) Verkleij, A. J.; Zwaal, R. F. A.; Roelofsen, B.; Comfurius, P.; Kastelijn, D.; van Deenen, L. L. M. The Asymmetric Distribution of Phospholipids in the Human Red Cell Membrane. A Combined Study Using Phospholipases and Freeze-Etch Electron Microscopy. *Biochimica et Biophysica Acta (BBA) - Biomembranes* **1973**, *323* (2), 178–193. https://doi.org/10.1016/0005-2736(73)90143-0.

(22) Vahedi, A.; Bigdelou, P.; Farnoud, A. M. Quantitative Analysis of Red Blood Cell Membrane Phospholipids and Modulation of Cell-Macrophage Interactions Using Cyclodextrins. *Scientific Reports* **2020**, *10* (1), 15111. https://doi.org/10.1038/s41598-020-72176-3.

(23) Liu, J.; Conboy, J. C. 1,2-Diacyl-Phosphatidylcholine Flip-Flop Measured Directly by Sum-Frequency Vibrational Spectroscopy. *Biophysical Journal* **2005**, *89* (4), 2522–2532. https://doi.org/10.1529/biophysj.105.065672.

(24) Marquardt, D.; Kučerka, N.; Wassall, S. R.; Harroun, T. A.; Katsaras, J. Cholesterol's Location in Lipid Bilayers. *Chemistry and Physics of Lipids* **2016**, *199*, 17–25. https://doi.org/10.1016/j.chemphyslip.2016.04.001.

(25) Ramstedt, B.; Slotte, J. P. Sphingolipids and the Formation of Sterol-Enriched Ordered Membrane Domains. *Biochimica et Biophysica Acta (BBA) - Biomembranes* **2006**, *1758* (12), 1945–1956. https://doi.org/10.1016/j.bbamem.2006.05.020.

(26) Tsamaloukas, A.; Szadkowska, H.; Heerklotz, H. Thermodynamic Comparison of the Interactions of Cholesterol with Unsaturated Phospholipid and Sphingomyelins. *Biophysical Journal* **2006**, *90* (12), 4479–4487. https://doi.org/10.1529/biophysj.105.080127.

(27) Devaux, P. F.; Morris, R. Transmembrane Asymmetry and Lateral Domains in Biological Membranes. *Traffic* **2004**, *5* (4), 241–246. https://doi.org/10.1111/j.1600-0854.2004.0170.x.

(28) Wang, W.; Yang, L.; Huang, H. W. Evidence of Cholesterol Accumulated in High Curvature Regions: Implication to the Curvature Elastic Energy for Lipid

Mixtures. *Biophysical Journal* **2007**, *92* (8), 2819–2830. https://doi.org/10.1529/biophysj.106.097923.

(29) Giang, H.; Schick, M. How Cholesterol Could Be Drawn to the Cytoplasmic Leaf of the Plasma Membrane by Phosphatidylethanolamine. *Biophysical Journal* **2014**, *107* (10), 2337–2344. https://doi.org/10.1016/j.bpj.2014.10.012.

(30) Courtney, K. C.; Pezeshkian, W.; Raghupathy, R.; Zhang, C.; Darbyson, A.; Ipsen, J. H.; Ford, D. A.; Khandelia, H.; Presley, J. F.; Zha, X. C24 Sphingolipids Govern the Transbilayer Asymmetry of Cholesterol and Lateral Organization of Model and Live-Cell Plasma Membranes. *Cell Reports* **2018**, *24* (4), 1037–1049. https://doi.org/10.1016/j.celrep.2018.06.104.

(31) Lange, Y.; Slayton, J. M. Interaction of Cholesterol and Lysophosphatidylcholine in Determining Red Cell Shape. *J. Lipid Res.* **1982**, *23* (8), 1121–1127.

(32) John, K.; Schreiber, S.; Kubelt, J.; Herrmann, A.; Müller, P. Transbilayer Movement of Phospholipids at the Main Phase Transition of Lipid Membranes: Implications for Rapid Flip-Flop in Biological Membranes. *Biophysical Journal* **2002**, *83* (6), 3315–3323. https://doi.org/10.1016/S0006-3495(02)75332-0.

(33) Leidl, K.; Liebisch, G.; Richter, D.; Schmitz, G. Mass Spectrometric Analysis of Lipid Species of Human Circulating Blood Cells. *Biochimica et Biophysica Acta (BBA) - Molecular and Cell Biology of Lipids* **2008**, *1781* (10), 655–664. https://doi.org/10.1016/j.bbalip.2008.07.008.

(34) Zehethofer, N.; Bermbach, S.; Hagner, S.; Garn, H.; Müller, J.; Goldmann, T.; Lindner, B.; Schwudke, D.; König, P. Lipid Analysis of Airway Epithelial Cells for Studying Respiratory Diseases. *Chromatographia* **2015**, *78* (5–6), 403–413. https://doi.org/10.1007/s10337-014-2787-5.

(35) Steck, T. L.; Lange, Y. Transverse Distribution of Plasma Membrane Bilayer Cholesterol: Picking Sides. *Traffic* **2018**, *19* (10), 750–760. https://doi.org/10.1111/tra.12586.

(36) Garg, S.; Porcar, L.; Woodka, A. C.; Butler, P. D.; Perez-Salas, U. Noninvasive Neutron Scattering Measurements Reveal Slower Cholesterol Transport in Model Lipid Membranes. *Biophysical Journal* **2011**, *101* (2), 370–377. https://doi.org/10.1016/j.bpj.2011.06.014.

(37) Bennett, W. F. D.; MacCallum, J. L.; Hinner, M. J.; Marrink, S. J.; Tieleman, D. P. Molecular View of Cholesterol Flip-Flop and Chemical Potential in Different Membrane Environments. *J. Am. Chem. Soc.* **2009**, *131* (35), 12714–12720. https://doi.org/10.1021/ja903529f.

(38) Jo, S.; Rui, H.; Lim, J. B.; Klauda, J. B.; Im, W. Cholesterol Flip-Flop: Insights from Free Energy Simulation Studies. *J. Phys. Chem. B* **2010**, *114* (42), 13342–13348. https://doi.org/10.1021/jp108166k.

(39) Steck, T. L.; Ye, J.; Lange, Y. Probing Red Cell Membrane Cholesterol Movement with Cyclodextrin. *Biophys J* **2002**, *83* (4), 2118–2125.

(40) Bruckner, R. J.; Mansy, S. S.; Ricardo, A.; Mahadevan, L.; Szostak, J. W. Flip-Flop-Induced Relaxation of Bending Energy: Implications for Membrane Remodeling. *Biophys J* **2009**, *97* (12), 3113–3122. https://doi.org/10.1016/j.bpj.2009.09.025.

(41) Vahedi, A.; Farnoud, A. M. Cyclodextrins for Probing Plasma Membrane Lipids. In *Analysis of Membrane Lipids*; Prasad, R., Singh, A., Eds.; Springer Protocols Handbooks; Springer US: New York, NY, 2020; pp 143–160. https://doi.org/10.1007/978-1-0716-0631-5_9.

(42) Yesylevskyy, S. O.; Schäfer, L. V.; Sengupta, D.; Marrink, S. J. Polarizable Water Model for the Coarse-Grained MARTINI Force Field. *PLOS Computational Biology* **2010**, *6* (6), e1000810. https://doi.org/10.1371/journal.pcbi.1000810.

(43) Michalowsky, J.; Schäfer, L. V.; Holm, C.; Smiatek, J. A Refined Polarizable Water Model for the Coarse-Grained MARTINI Force Field with Long-Range Electrostatic Interactions. *The Journal of Chemical Physics* **2017**, *146* (5), 054501. https://doi.org/10.1063/1.4974833.

(44) Marrink, S. J.; Risselada, H. J.; Yefimov, S.; Tieleman, D. P.; de Vries, A. H. The MARTINI Force Field: Coarse Grained Model for Biomolecular Simulations. *The Journal of Physical Chemistry B* **2007**, *111* (27), 7812–7824. https://doi.org/10.1021/jp071097f.

(45) Wassenaar, T. A.; Ingólfsson, H. I.; Böckmann, R. A.; Tieleman, D. P.; Marrink, S. J. Computational Lipidomics with Insane: A Versatile Tool for Generating Custom Membranes for Molecular Simulations. *Journal of chemical theory and computation* **2015**, *11* (5), 2144–2155.

(46) Abraham, M. J.; Murtola, T.; Schulz, R.; Páll, S.; Smith, J. C.; Hess, B.; Lindahl, E. GROMACS: High Performance Molecular Simulations through Multi-Level Parallelism from Laptops to Supercomputers. *SoftwareX* **2015**, *1–2*, 19–25. https://doi.org/10.1016/j.softx.2015.06.001.

(47) Bussi, G.; Donadio, D.; Parrinello, M. Canonical Sampling through Velocity-Rescaling. *The Journal of Chemical Physics* **2007**, *126* (1), 014101. https://doi.org/10.1063/1.2408420.

(48) Huang, C.-H. Roles of Carbonyl Oxygens at the Bilayer Interface in Phospholipid–Sterol Interaction. *Nature* **1976**, *259* (5540), 242–244. https://doi.org/10.1038/259242a0.

(49) Oh, Y.; Sung, B. J. Facilitated and Non-Gaussian Diffusion of Cholesterol in Liquid Ordered Phase Bilayers Depends on the Flip-Flop and Spatial Arrangement of Cholesterol. *J. Phys. Chem. Lett.* **2018**, 6529–6535. https://doi.org/10.1021/acs.jpclett.8b02982.

(50) Zhang, Y.; Lervik, A.; Seddon, J.; Bresme, F. A Coarse-Grained Molecular Dynamics Investigation of the Phase Behavior of DPPC/Cholesterol Mixtures. *Chemistry and Physics of Lipids* **2015**, *185*, 88–98. https://doi.org/10.1016/j.chemphyslip.2014.07.011.

(51) Bhide, S. Y.; Zhang, Z.; Berkowitz, M. L. Molecular Dynamics Simulations of SOPS and Sphingomyelin Bilayers Containing Cholesterol. *Biophysical Journal* **2007**, *92* (4), 1284–1295. https://doi.org/10.1529/biophysj.106.096214.

(52) Arnarez, C. ment; Uusitalo, J. Dry Martini, a Coarse-Grained Force Field for Lipid Membrane Simulations with Implicit Solvent. *JCTC* **2014**.

(53) Krause, M. R.; Regen, S. L. The Structural Role of Cholesterol in Cell Membranes: From Condensed Bilayers to Lipid Rafts. *Acc. Chem. Res.* **2014**, *47* (12), 3512–3521. https://doi.org/10.1021/ar500260t.

(54) Huang, J.; Buboltz, J. T.; Feigenson, G. W. Maximum Solubility of Cholesterol in Phosphatidylcholine and Phosphatidylethanolamine Bilayers. *Biochimica et Biophysica Acta (BBA) - Biomembranes* **1999**, *1417* (1), 89–100. https://doi.org/10.1016/S0005-2736(98)00260-0.

(55) Marrink, S. J.; de Vries, A. H.; Harroun, Thad. A.; Katsaras, J.; Wassall, S. R. Cholesterol Shows Preference for the Interior of Polyunsaturated Lipid Membranes. *J. Am. Chem. Soc.* **2008**, *130* (1), 10–11. https://doi.org/10.1021/ja076641c.

(56) García-Arribas, A. B.; Alonso, A.; Goñi, F. M. Cholesterol Interactions with Ceramide and Sphingomyelin. *Chemistry and Physics of Lipids* **2016**, *199*, 26–34. https://doi.org/10.1016/j.chemphyslip.2016.04.002.

(57) Chiantia, S.; London, E. Acyl Chain Length and Saturation Modulate Interleaflet Coupling in Asymmetric Bilayers: Effects on Dynamics and Structural Order. *Biophysical Journal* **2012**, *103* (11), 2311–2319. https://doi.org/10.1016/j.bpj.2012.10.033.

(58) Nagle, J. F.; Tristram-Nagle, S. Structure of Lipid Bilayers. *Biochimica et Biophysica Acta (BBA) - Reviews on Biomembranes* **2000**, *1469* (3), 159–195. https://doi.org/10.1016/S0304-4157(00)00016-2.

(59) Leventis, R.; Silvius, J. R. Use of Cyclodextrins to Monitor Transbilayer Movement and Differential Lipid Affinities of Cholesterol. *Biophysical Journal* **2001**, *81* (4), 2257–2267. https://doi.org/10.1016/S0006-3495(01)75873-0.

(60) Backer, J. M.; Dawidowicz, E. A. Transmembrane Movement of Cholesterol in Small Unilamellar Vesicles Detected by Cholesterol Oxidase. *J. Biol. Chem.* **1981**, *256* (2), 586–588.

(61) Steck, T. L.; Lange, Y. How Slow Is the Transbilayer Diffusion (Flip-Flop) of Cholesterol? *Biophysical Journal* **2012**, *102* (4), 945–946. https://doi.org/10.1016/j.bpj.2011.10.059.

(62) Kucerka, N.; Nieh, M.-P.; Pencer, J.; Sachs, J. N.; Katsaras, J. What Determines the Thickness of a Biological Membrane? *Gen. Physiol. Biophys.* **2009**, *28* (2), 117–125. https://doi.org/10.4149/gpb_2009_02_117.

(63) Martinez-Seara, H.; Róg, T.; Karttunen, M.; Vattulainen, I.; Reigada, R. Cholesterol Induces Specific Spatial and Orientational Order in Cholesterol/Phospholipid Membranes. *PLOS ONE* **2010**, *5* (6), e11162. https://doi.org/10.1371/journal.pone.0011162.

(64) Dai, J.; Alwarawrah, M.; Huang, J. Instability Of Cholesterol Clusters In Lipid Bilayers And The Cholesterol's Umbrella Effect. *J Phys Chem B* **2010**, *114* (2), 840. https://doi.org/10.1021/jp909061h.

(65) Yesylevskyy, S. O.; Demchenko, A. P. How Cholesterol Is Distributed between Monolayers in Asymmetric Lipid Membranes. *European Biophysics Journal* **2012**, *41* (12), 1043–1054. https://doi.org/10.1007/s00249-012-0863-z.

(66) Ali, M. R.; Cheng, K. H.; Huang, J. Assess the Nature of Cholesterol–Lipid Interactions through the Chemical Potential of Cholesterol in Phosphatidylcholine Bilayers. *PNAS* **2007**, *104* (13), 5372–5377. https://doi.org/10.1073/pnas.0611450104.

(67) Lin, Q.; London, E. Preparation of Artificial Plasma Membrane Mimicking Vesicles with Lipid Asymmetry. *PLOS ONE* **2014**, *9* (1), e87903. https://doi.org/10.1371/journal.pone.0087903.

(68) Edholm, O.; Nagle, J. F. Areas of Molecules in Membranes Consisting of Mixtures. *Biophysical Journal* **2005**, *89* (3), 1827–1832. https://doi.org/10.1529/biophysj.105.064329.

(69) Kučerka, N.; Nieh, M.-P.; Katsaras, J. Fluid Phase Lipid Areas and Bilayer Thicknesses of Commonly Used Phosphatidylcholines as a Function of Temperature. *Biochimica et Biophysica Acta (BBA) - Biomembranes* **2011**, *1808* (11), 2761–2771. https://doi.org/10.1016/j.bbamem.2011.07.022.

(70) Wang, Y.; Gkeka, P.; Fuchs, J. E.; Liedl, K. R.; Cournia, Z. DPPC-Cholesterol Phase Diagram Using Coarse-Grained Molecular Dynamics Simulations. *Biochimica et Biophysica Acta (BBA) - Biomembranes* **2016**, *1858* (11), 2846–2857. https://doi.org/10.1016/j.bbamem.2016.08.005.

(71) Lee, A. G. How Lipids Affect the Activities of Integral Membrane Proteins. *Biochimica et Biophysica Acta (BBA) - Biomembranes* **2004**, *1666* (1), 62–87. https://doi.org/10.1016/j.bbamem.2004.05.012.

(72) Pan, J.; Mills, T. T.; Tristram-Nagle, S.; Nagle, J. F. Cholesterol Perturbs Lipid Bilayers Nonuniversally. *Phys. Rev. Lett.* **2008**, *100* (19), 198103. https://doi.org/10.1103/PhysRevLett.100.198103.

(73) Feigenson, G. W. Phase Diagrams and Lipid Domains in Multicomponent Lipid Bilayer Mixtures. *Biochimica et Biophysica Acta (BBA) - Biomembranes* **2009**, *1788* (1), 47–52. https://doi.org/10.1016/j.bbamem.2008.08.014.

(74) Armstrong, C. L.; Marquardt, D.; Dies, H.; Kučerka, N.; Yamani, Z.; Harroun, T. A.; Katsaras, J.; Shi, A.-C.; Rheinstädter, M. C. The Observation of Highly Ordered Domains in Membranes with Cholesterol. *PLOS ONE* **2013**, *8* (6), e66162. https://doi.org/10.1371/journal.pone.0066162.

(75) Bagatolli, L.; Kumar, P. B. S. Phase Behavior of Multicomponent Membranes: Experimental and Computational Techniques. *Soft Matter* **2009**, *5* (17), 3234–3248. https://doi.org/10.1039/B901866B.

(76) Zhuang, X.; Makover, J. R.; Im, W.; Klauda, J. B. A Systematic Molecular Dynamics Simulation Study of Temperature Dependent Bilayer Structural Properties. *Biochimica et Biophysica Acta (BBA) - Biomembranes* **2014**, *1838* (10), 2520–2529. https://doi.org/10.1016/j.bbamem.2014.06.010.

(77) Marquardt, D.; Kučerka, N.; Wassall, S. R.; Harroun, T. A.; Katsaras, J. Cholesterol's Location in Lipid Bilayers. *Chemistry and Physics of Lipids* **2016**, *199*, 17–25. https://doi.org/10.1016/j.chemphyslip.2016.04.001.

(78) Khakbaz, P.; Klauda, J. B. Investigation of Phase Transitions of Saturated Phosphocholine Lipid Bilayers via Molecular Dynamics Simulations. *Biochimica*

et Biophysica Acta (BBA) - Biomembranes **2018**, *1860* (8), 1489–1501.
https://doi.org/10.1016/j.bbamem.2018.04.014.

(79) Konyakhina, T. M.; Feigenson, G. W. Phase Diagram of a Polyunsaturated Lipid
Mixture: Brain Sphingomyelin/1-Stearoyl-2-Docosahexaenoyl-Sn-Glycero-3-
Phosphocholine/Cholesterol. *Biochimica et Biophysica Acta (BBA) -
Biomembranes* **2016**, *1858* (1), 153–161.
https://doi.org/10.1016/j.bbamem.2015.10.016.

(80) Konyakhina, T. M.; Wu, J.; Mastroianni, J. D.; Heberle, F. A.; Feigenson, G. W.
Phase Diagram of a 4-Component Lipid Mixture: DSPC/DOPC/POPC/Chol.
Biochimica et Biophysica Acta (BBA) - Biomembranes **2013**, *1828* (9), 2204–2214.
https://doi.org/10.1016/j.bbamem.2013.05.020.

(81) Ueda, M. J.; Ito, T.; Okada, T. S.; Ohnishi, S. I. A Correlation between Membrane
Fluidity and the Critical Temperature for Cell Adhesion. *The Journal of Cell
Biology* **1976**, *71* (2), 670–674. https://doi.org/10.1083/jcb.71.2.670.

(82) Marquardt, D.; Heberle, F. A.; Nickels, J. D.; Pabst, G.; Katsaras, J. On Scattered
Waves and Lipid Domains: Detecting Membrane Rafts with X-Rays and Neutrons.
Soft Matter **2015**, *11* (47), 9055–9072. https://doi.org/10.1039/C5SM01807B.

(83) Buboltz, J. T.; Feigenson, G. W. A Novel Strategy for the Preparation of
Liposomes: Rapid Solvent Exchange. *Biochimica et Biophysica Acta (BBA) -
Biomembranes* **1999**, *1417* (2), 232–245. https://doi.org/10.1016/S0005-
2736(99)00006-1.

(84) Feigenson, G. W.; Buboltz, J. T. Ternary Phase Diagram of Dipalmitoyl-
PC/Dilauroyl-PC/Cholesterol: Nanoscopic Domain Formation Driven by
Cholesterol. *Biophysical Journal; New York* **2001**, *80* (6), 2775–2788.
https://doi.org/10.1016/S0006-3495(01)76245-5.

(85) Harroun, T. A.; Katsaras, J.; Wassall, S. R. Cholesterol Is Found To Reside in the
Center of a Polyunsaturated Lipid Membrane. *Biochemistry* **2008**, *47* (27), 7090–
7096. https://doi.org/10.1021/bi800123b.

(86) Simons, K.; Toomre, D. Lipid Rafts and Signal Transduction. *Nature Reviews
Molecular Cell Biology* **2000**, *1* (1), 31–39. https://doi.org/10.1038/35036052.

(87) Davis, J. H.; Clair, J. J.; Juhasz, J. Phase Equilibria in DOPC/DPPC-
D62/Cholesterol Mixtures. *Biophysical Journal* **2009**, *96* (2), 521–539.
https://doi.org/10.1016/j.bpj.2008.09.042.

(88) Li, X.-M.; Smaby, J. M.; Momsen, M. M.; Brockman, H. L.; Brown, R. E.
Sphingomyelin Interfacial Behavior: The Impact of Changing Acyl Chain
Composition. *Biophysical Journal* **2000**, *78* (4), 1921–1931.
https://doi.org/10.1016/S0006-3495(00)76740-3.

(89) Petrache, H. I.; Tristram-Nagle, S.; Gawrisch, K.; Harries, D.; Parsegian, V. A.;
Nagle, J. F. Structure and Fluctuations of Charged Phosphatidylserine Bilayers in
the Absence of Salt. *Biophysical Journal* **2004**, *86* (3), 1574–1586.
https://doi.org/10.1016/S0006-3495(04)74225-3.

(90) Liu, A.; Qi, X. Molecular Dynamics Simulations of DOPC Lipid Bilayers: The
Effect of Lennard-Jones Parameters of Hydrocarbon Chains. *Comput Mol Biosci*
2012, *2* (3), 78–82. https://doi.org/10.4236/cmb.2012.23007.

(91) Bhide, S. Y.; Berkowitz, M. L. Structure and Dynamics of Water at the Interface with Phospholipid Bilayers. *The Journal of Chemical Physics* **2005**, *123* (22), 224702. https://doi.org/10.1063/1.2132277.

(92) Polyansky, A. A.; Volynsky, P. E.; Nolde, D. E.; Arseniev, A. S.; Efremov, R. G. Role of Lipid Charge in Organization of Water/Lipid Bilayer Interface: Insights via Computer Simulations. *J. Phys. Chem. B* **2005**, *109* (31), 15052–15059. https://doi.org/10.1021/jp0510185.

(93) Maulik, P. R.; Shipley, G. G. N-Palmitoyl Sphingomyelin Bilayers: Structure and Interactions with Cholesterol and Dipalmitoylphosphatidylcholine. *Biochemistry* **1996**, *35* (24), 8025–8034. https://doi.org/10.1021/bi9528356.

(94) Niemelä, P.; Hyvönen, M. T.; Vattulainen, I. Structure and Dynamics of Sphingomyelin Bilayer: Insight Gained through Systematic Comparison to Phosphatidylcholine. *Biophys J* **2004**, *87* (5), 2976–2989. https://doi.org/10.1529/biophysj.104.048702.

(95) Almeida, R. F. M. de; Fedorov, A.; Prieto, M. Sphingomyelin/Phosphatidylcholine/Cholesterol Phase Diagram: Boundaries and Composition of Lipid Rafts. *Biophysical Journal; New York* **2003**, *85* (4), 2406–2416. https://doi.org/10.1016/S0006-3495(03)74664-5.

(96) Aghaaminiha, M.; Sharma, S. Spatial Distribution of Cholesterol in Lipid Bilayers. *bioRxiv* **2019**, 636845. https://doi.org/10.1101/636845.

(97) Aghaaminiha, M.; Sharma, S. Chicken-Or-Egg Conundrum: Ordering of Lipids Vs. Cholesterol Localization in the Asymmetric Lipid Bilayers. *Biophysical Journal* **2021**, *120* (3), 147a. https://doi.org/10.1016/j.bpj.2020.11.1081.

(98) Aghaaminiha (Amini), M. (Reza); Sharma, S. Cholesterol Spatial Distribution in Asymmetric Lipid Bilayers. *Biophysical Journal* **2020**, *118* (3), 386a–387a. https://doi.org/10.1016/j.bpj.2019.11.2203.

(99) Aghaaminiha, M.; Ghanadian, S. A.; Ahmadi, E.; Farnoud, A. M. A Machine Learning Approach to Estimation of Phase Diagrams for Three-Component Lipid Mixtures. *Biochimica et Biophysica Acta (BBA) - Biomembranes* **2020**, *1862* (9), 183350. https://doi.org/10.1016/j.bbamem.2020.183350.

(100) Elson, E. L.; Fried, E.; Dolbow, J. E.; Genin, G. M. Phase Separation in Biological Membranes: Integration of Theory and Experiment. *Annual Review of Biophysics* **2010**, *39* (1), 207–226. https://doi.org/10.1146/annurev.biophys.093008.131238.

(101) Farnoud, A. M.; Toledo, A. M.; Konopka, J. B.; Del Poeta, M.; London, E. Raft-like Membrane Domains in Pathogenic Microorganisms. In *Current Topics in Membranes*; Elsevier, 2015; Vol. 75, pp 233–268. https://doi.org/10.1016/bs.ctm.2015.03.005.

(102) Marsh, D. Cholesterol-Induced Fluid Membrane Domains: A Compendium of Lipid-Raft Ternary Phase Diagrams. *Biochimica et Biophysica Acta (BBA) - Biomembranes* **2009**, *1788* (10), 2114–2123. https://doi.org/10.1016/j.bbamem.2009.08.004.

(103) Parton, R. G.; Richards, A. A. Lipid Rafts and Caveolae Asportals for Endocytosis: New Insights and Common Mechanisms. *Traffic* **2003**, *4* (11), 724–738. https://doi.org/10.1034/j.1600-0854.2003.00128.x.

(104) Alvarez, F. J.; Douglas, L. M.; Konopka, J. B. Sterol-Rich Plasma Membrane Domains in Fungi. *Eukaryotic Cell* **2007**, *6* (5), 755–763. https://doi.org/10.1128/EC.00008-07.

(105) Uppamoochikkal, P.; Tristram-Nagle, S.; Nagle, J. F. Orientation of Tie-Lines in the Phase Diagram of DOPC/DPPC/Cholesterol Model Biomembranes. *Langmuir* **2010**, *26* (22), 17363–17368. https://doi.org/10.1021/la103024f.

(106) Cevc, G. How Membrane Chain-Melting Phase-Transition Temperature Is Affected by the Lipid Chain Asymmetry and Degree of Unsaturation: An Effective Chain-Length Model. *Biochemistry* **1991**, *30* (29), 7186–7193. https://doi.org/10.1021/bi00243a021.

(107) Wang, L.; Irausquin, S. J.; Yang, J. Y. Prediction of Lipid-Interacting Amino Acid Residues from Sequence Features. *Int J Comput Biol Drug Des* **2008**, *1* (1), 14–25. https://doi.org/10.1504/ijcbdd.2008.018707.

(108) Cho, H.; Wu, M.; Bilgin, B.; Walton, S. P.; Chan, C. Latest Developments in Experimental and Computational Approaches to Characterize Protein–Lipid Interactions. *PROTEOMICS* **2012**, *12* (22), 3273–3285. https://doi.org/10.1002/pmic.201200255.

(109) Löpez, C. A.; Vesselinov, V. V.; Gnanakaran, S.; Alexandrov, B. S. Unsupervised Machine Learning for Analysis of Phase Separation in Ternary Lipid Mixture. *J Chem Theory Comput* **2019**, *15* (11), 6343–6357. https://doi.org/10.1021/acs.jctc.9b00074.

(110) Mitra, E. D.; Whitehead, S. C.; Holowka, D.; Baird, B.; Sethna, J. P. Computation of a Theoretical Membrane Phase Diagram and the Role of Phase in Lipid-Raft-Mediated Protein Organization. *J. Phys. Chem. B* **2018**, *122* (13), 3500–3513. https://doi.org/10.1021/acs.jpcb.7b10695.

(111) Caffrey, M. *Lipid Thermotropic Phase Transition Database (LIPIDAT). User's Guide Version 1.0 Version 1.0*; U.S. Dept. of Commerce, National Institute of Standards and Technology, Standard Reference Data Program: Gaithersburg, MD, 1993.

(112) Silvius, J. R. *Thermotropic Phase Transitions of Pure Lipids in Model Membranes and Their Modifications by Membrane Proteins*; John Wiley & Sons: New York, 1982.

(113) Chawla, N. V.; Bowyer, K. W.; Hall, L. O.; Kegelmeyer, W. P. SMOTE: Synthetic Minority over-Sampling Technique. *1* **2002**, *16*, 321–357. https://doi.org/10.1613/jair.953.

(114) Lahmiri, S. A Comparative Study of Backpropagation Algorithms in Financial Prediction. *International Journal of Computer Science, Engineering and Applications (IJCSEA)* **2011**, *1* (4). https://doi.org/10.5121/ijcsea.2011.1402.

(115) Roy, R. K. *A Primer on the Taguchi Method*, 2nd ed.; Society of Manufacturing Engineers: Dearborn, MI, 2010.

(116) Mousavi, S. M.; Hajipour, V.; Niaki, S. T. A.; Alikar, N. Optimizing Multi-Item Multi-Period Inventory Control System with Discounted Cash Flow and Inflation: Two Calibrated Meta-Heuristic Algorithms. *Applied Mathematical Modelling* **2013**, *37* (4), 2241–2256. https://doi.org/10.1016/j.apm.2012.05.019.

(117) Sokolova, M.; Lapalme, G. A Systematic Analysis of Performance Measures for Classification Tasks. *Information Processing & Management* **2009**, *45* (4), 427–437. https://doi.org/10.1016/j.ipm.2009.03.002.

(118) Murray, S. M.; O'Brien, R. A.; Mattson, K. M.; Ceccarelli, C.; Sykora, R. E.; West, K. N.; Davis, J. H. The Fluid-Mosaic Model, Homeoviscous Adaptation, and Ionic Liquids: Dramatic Lowering of the Melting Point by Side-Chain Unsaturation. *Angewandte Chemie International Edition* **2010**, *49* (15), 2755–2758. https://doi.org/10.1002/anie.200906169.

(119) Cullis, P. R.; De Kruijff, B. The Polymorphic Phase Behaviour of Phosphatidylethanolamines of Natural and Synthetic Origin. A 31P NMR Study. *Biochimica et Biophysica Acta (BBA) - Biomembranes* **1978**, *513* (1), 31–42. https://doi.org/10.1016/0005-2736(78)90109-8.

(120) Castresana, J.; Nieva, J. L.; Rivas, E.; Alonso, A. Partial Dehydration of Phosphatidylethanolamine Phosphate Groups during Hexagonal Phase Formation, as Seen by i.r. Spectroscopy. *Biochem J* **1992**, *282* (Pt 2), 467–470.

(121) Martin, J. A.; Valone, F. W. The Existence of Imidazoline Corrosion Inhibitors. *Corrosion* **1985**, *41* (5), 281–287. https://doi.org/10.5006/1.3582003.

(122) McMahon, A. J. The Mechanism of Action of an Oleic Imidazoline Based Corrosion Inhibitor for Oilfield Use. *Colloids and Surfaces* **1991**, *59*, 187–208. https://doi.org/10.1016/0166-6622(91)80247-L.

(123) Edwards, A.; Osborne, C.; Webster, S.; Klenerman, D.; Joseph, M.; Ostovar, P.; Doyle, M. Mechanistic Studies of the Corrosion Inhibitor Oleic Imidazoline. *Corrosion Science* **1994**, *36* (2), 315–325. https://doi.org/10.1016/0010-938X(94)90160-0.

(124) Jaschke, M.; Butt, H.-J.; Gaub, H. E.; Manne, S. Surfactant Aggregates at a Metal Surface. *Langmuir* **1997**, *13* (6), 1381–1384. https://doi.org/10.1021/la9607767.

(125) Xiong, Y.; Brown, B.; Kinsella, B.; Nešić, S.; Pailleret, A. Atomic Force Microscopy Study of the Adsorption of Surfactant Corrosion Inhibitor Films. *Corrosion* **2013**, *70* (3), 247–260. https://doi.org/10.5006/0915.

(126) Ko, X.; Sharma, S. Adsorption and Self-Assembly of Surfactants on Metal–Water Interfaces. *J. Phys. Chem. B* **2017**, *121* (45), 10364–10370. https://doi.org/10.1021/acs.jpcb.7b09297.

(127) Finšgar, M.; Jackson, J. Application of Corrosion Inhibitors for Steels in Acidic Media for the Oil and Gas Industry: A Review. *Corrosion Science* **2014**, *86*, 17–41. https://doi.org/10.1016/j.corsci.2014.04.044.

(128) Pots, B. F. M. Mechanistic Models for the Prediction of CO2 Corrosion Rates under Multi-Phase Flow Conditions. **1995**.

(129) Zhang, R.; Gopal, M.; Jepson, W. P. Development of a Mechanistic Model for Predicting Corrosion Rate in Multiphase Oil/Water/Gas Flows. **1997**, 30.

(130) Anderko, A.; Young, R. D.; Plains, M. Simulation of CO2/H2S Corrosion Using Thermodynamic and Electrochemical Models. *NACE* **1999**, 19.

(131) Nesic, S.; Cai, J.; Wang, S.; Xiao, Y.; Liu, D. *Ohio University Multiphase Flow and Corrosion Prediction Software Package MULTICORP V4.0*; Ohio University, 2004.

(132) Nesic, S. Key Issues Related to Modelling of Internal Corrosion of Oil and Gas Pipelines – A Review. *Corrosion Science* **2007**, *49* (12), 4308–4338. https://doi.org/10.1016/j.corsci.2007.06.006.

(133) Gulbrandsen, E.; Nesic, S.; Stangeland, A.; Burchardtt, T. Effect of Precorrosion on the Performance of Inhibitors for CO2 Corrosion of Carbon Steel. *Corrosion* **1998**, *98*.

(134) Khodyrev, Yu. P.; Batyeva, E. S.; Badeeva, E. K.; Platova, E. V.; Tiwari, L.; Sinyashin, O. G. The Inhibition Action of Ammonium Salts of O,O′-Dialkyldithiophosphoric Acid on Carbon Dioxide Corrosion of Mild Steel. *Corrosion Science* **2011**, *53* (3), 976–983. https://doi.org/10.1016/j.corsci.2010.11.030.

(135) Rihan, R.; Shawabkeh, R.; Al-Bakr, N. The Effect of Two Amine-Based Corrosion Inhibitors in Improving the Corrosion Resistance of Carbon Steel in Sea Water. *J. of Materi Eng and Perform* **2014**, *23* (3), 693–699. https://doi.org/10.1007/s11665-013-0790-x.

(136) Javidi, M.; Chamanfar, R.; Bekhrad, S. Investigation on the Efficiency of Corrosion Inhibitor in CO2 Corrosion of Carbon Steel in the Presence of Iron Carbonate Scale. *Journal of Natural Gas Science and Engineering* **2019**, *61*, 197–205. https://doi.org/10.1016/j.jngse.2018.11.017.

(137) Xia, S.; Qiu, M.; Yu, L.; Liu, F.; Zhao, H. Molecular Dynamics and Density Functional Theory Study on Relationship between Structure of Imidazoline Derivatives and Inhibition Performance. *Corrosion Science* **2008**, *50* (7), 2021–2029. https://doi.org/10.1016/j.corsci.2008.04.021.

(138) Kokalj, A. Formation and Structure of Inhibitive Molecular Film of Imidazole on Iron Surface. *Corrosion Science* **2013**, *68*, 195–203. https://doi.org/10.1016/j.corsci.2012.11.015.

(139) Kurapati, Y.; Sharma, S. Adsorption Free Energies of Imidazolinium-Type Surfactants in Infinite Dilution and in Micellar State on Gold Surface. *J. Phys. Chem. B* **2018**, *122* (22), 5933–5939. https://doi.org/10.1021/acs.jpcb.8b02358.

(140) Sharma, S.; Ko, X.; Kurapati, Y.; Singh, H.; Nesic, S. Adsorption Behavior of Organic Corrosion Inhibitors on Metal Surfaces—Some New Insights from Molecular Simulations. *Corrosion* **2018**, *75* (1), 90–105. https://doi.org/10.5006/2976.

(141) Singh, H.; Sharma, S. Disintegration of Surfactant Micelles at Metal–Water Interfaces Promotes Their Strong Adsorption. *J. Phys. Chem. B* **2020**, *124* (11), 2262–2267. https://doi.org/10.1021/acs.jpcb.9b10780.

(142) Singh, H.; Sharma, S. Free Energy Profiles of Adsorption of Surfactant Micelles at Metal-Water Interfaces. *Molecular Simulation* **2020**, *0* (0), 1–8. https://doi.org/10.1080/08927022.2020.1780231.

(143) Nesic, S.; Nordsveen, M.; Maxwell, N.; Vrhovac, M. Probabilistic Modelling of CO2 Corrosion Laboratory Data Using Neural Networks. *Corrosion Science* **2001**, *43* (7), 1373–1392. https://doi.org/10.1016/S0010-938X(00)00157-8.

(144) Nash, W. T.; Powell, C. J.; Drummond, T.; Birbilis, N. Automated Corrosion Detection Using Crowdsourced Training for Deep Learning. *Corrosion* **2019**, *76* (2), 135–141. https://doi.org/10.5006/3397.

(145) Ossai, C. I. A Data-Driven Machine Learning Approach for Corrosion Risk Assessment—A Comparative Study. *Big Data and Cognitive Computing* **2019**, *3* (2), 28. https://doi.org/10.3390/bdcc3020028.

(146) Sanchez, G.; Aperador, W.; Cerón, A. Corrosion Grade Classification: A Machine Learning Approach. *Indian Chemical Engineer* **2020**, *62* (3), 277–286. https://doi.org/10.1080/00194506.2019.1675539.

(147) Yan, L.; Diao, Y.; Lang, Z.; Gao, K. Corrosion Rate Prediction and Influencing Factors Evaluation of Low-Alloy Steels in Marine Atmosphere Using Machine Learning Approach. *Science and Technology of Advanced Materials* **2020**, *21* (1), 359–370. https://doi.org/10.1080/14686996.2020.1746196.

(148) Salami, B. A.; Rahman, S. M.; Oyehan, T. A.; Maslehuddin, M.; Al Dulaijan, S. U. Ensemble Machine Learning Model for Corrosion Initiation Time Estimation of Embedded Steel Reinforced Self-Compacting Concrete. *Measurement* **2020**, *165*, 108141. https://doi.org/10.1016/j.measurement.2020.108141.

(149) Gong, X.; Dong, C.; Xu, J.; Wang, L.; Li, X. Machine Learning Assistance for Electrochemical Curve Simulation of Corrosion and Its Application. *Materials and Corrosion* **2020**, *71* (3), 474–484. https://doi.org/10.1002/maco.201911224.

(150) Aulia, R.; Tan, H.; Sriramula, S. Prediction of Corroded Pipeline Performance Based on Dynamic Reliability Models. *Procedia CIRP* **2019**, *80*, 518–523. https://doi.org/10.1016/j.procir.2019.01.093.

(151) Abass, A.; Wada, K.; Matsunaga, H.; Remes, H.; Vuorio, T. Quantitative Characterization of the Spatial Distribution of Corrosion Pits Based on Nearest Neighbor Analysis. *Corrosion* **2020**, *76* (9), 861–870. https://doi.org/10.5006/3551.

(152) Famili, A.; Shen, W.-M.; Weber, R.; Simoudis, E. Data Preprocessing and Intelligent Data Analysis. *Intelligent Data Analysis* **1997**, *1* (1), 3–23. https://doi.org/10.3233/IDA-1997-1102.

(153) Kotsiantis, S. B.; Kanellopoulos, D.; Pintelas, P. E. Data Preprocessing for Supervised Leaning. *IJCS* **2006**, *1* (1), 111–117.

(154) Murakawa, T.; Nagaura, S.; Hackerman, N. Coverage of Iron Surface by Organic Compounds and Anions in Acid Solutions. *Corrosion Science* **1967**, *7* (2), 79–89. https://doi.org/10.1016/S0010-938X(67)80105-7.

(155) Marrink, S.; Risselada, J. The MARTINI Force Field: Coarse Grained Model for Biomolecular Simulations. **2007**.

(156) Saiz, L.; Klein, M. L. Influence of Highly Polyunsaturated Lipid Acyl Chains of Biomembranes on the NMR Order Parameters. *J. Am. Chem. Soc.* **2001**, *123* (30), 7381–7387. https://doi.org/10.1021/ja003987d.

(157) Arsov, Z.; González-Ramírez, E. J.; Goñi, F. M.; Tristram-Nagle, S.; Nagle, J. F. Phase Behavior of Palmitoyl and Egg Sphingomyelin. *Chemistry and Physics of Lipids* **2018**, *213*, 102–110. https://doi.org/10.1016/j.chemphyslip.2018.03.003.

(158) Lopez, C.; Cheng, K.; Perez, J. Thermotropic Phase Behavior of Milk Sphingomyelin and Role of Cholesterol in the Formation of the Liquid Ordered Phase Examined Using SR-XRD and DSC. *Chem. Phys. Lipids* **2018**, *215*, 46–55. https://doi.org/10.1016/j.chemphyslip.2018.07.008.

Appendix A: Correspond to the Materials Presented in Chapter 3

Table A1. Various properties of the three studied bilayers. Cholesterol was present in all the bilayers in the molar ratio of 1:2 with respect to the total number of lipids. The values of area per lipid is smaller compared to the usually reported values (~ 0.5 to 0.7 nm^2) for the bilayers where no cholesterol present.[58]

Lipid bilayer	System (box) area (nm^2)	Thickness ($\mathrm{\AA}$)	Area per lipid (nm^2)		Average order parameter (S_{chain})	
			Inner	Outer	Inner	Outer
Asymmetric	507.63 ± 0.07	41.24 ± 0.14	0.52 ± 0.01	0.48 ± 0.01	0.33 ± 0.02	0.48 ± 0.02
Cyto-symmetric	530.43 ± 0.10	40.28 ± 0.14	0.52 ± 0.01	0.52 ± 0.01	0.32 ± 0.02	0.32 ± 0.02
Exo-symmetric	484.77 ± 0.12	42.25 ± 0.14	0.47 ± 0.01	0.47 ± 0.01	0.48 ± 0.02	0.48 ± 0.02

Table A2. Properties of equimolar asymmetric bilayer systems with DPPC as the outer-leaflet. Cholesterol was present in all the bilayers in the molar ratio of 1:2 with respect to the total number of lipids. The "i" and "o" subscripts represent the inner and outer leaflets, respectively.

Lipid bilayer	System (box) area (nm^2)	Thickness (Å)	Flip-flop rate 1×10^6 1/s	Area per lipid (nm^2) Inner	Area per lipid (nm^2) Outer
DPPCi/DPPCo/CHOL	114.71 ± 0.84	45.11 ± 0.25	0.26 ± 0.06	0.45 ± 0.01	0.45 ± 0.01
16:0SMi/DPPCo/CHOL	115.28 ± 0.93	43.52 ± 0.25	0.36 ± 0.06	0.44 ± 0.01	0.46 ± 0.01
DPPEi/DPPCo/CHOL	114.39 ± 0.80	45.17 ± 0.24	0.22 ± 0.04	0.45 ± 0.01	0.45 ± 0.01
DPPSi/DPPCo/CHOL	115.59 ± 0.90	44.97 ± 0.25	0.29 ± 0.05	0.45 ± 0.01	0.46 ± 0.01
DOPCi/DPPCo/CHOL	124.60 ± 1.07	42.62 ± 0.28	2.40 ± 0.32	0.53 ± 0.01	0.45 ± 0.01
POPCi/DPPCo/CHOL	120.50 ± 0.98	43.60 ± 0.25	1.30 ± 0.17	0.50 ± 0.01	0.44 ± 0.01
12:0PCi/DPPCo/CHOL	115.36 ± 0.86	41.81 ± 0.21	0.52 ± 0.07	0.45 ± 0.01	0.45 ± 0.01
24:0PCi/DPPCo/CHOL	114.10 ± 0.93	51.69 ± 0.27	2.10 ± 0.18	0.43 ± 0.01	0.46 ± 0.01
DOPEi/DPPCo/CHOL	123.61 ± 1.04	42.78 ± 0.26	2.10 ± 0.24	0.53 ± 0.01	0.45 ± 0.01
POPEi/DPPCo/CHOL	119.64 ± 0.93	43.75 ± 0.25	1.10 ± 0.17	0.49 ± 0.01	0.45 ± 0.01
DOPSi/DPPCo/CHOL	125.58 ± 1.11	42.42 ± 0.29	2.20 ± 0.25	0.53 ± 0.01	0.46 ± 0.01
POPSi/DPPCo/CHOL	121.77 ± 1.06	43.39 ± 0.26	1.30 ± 0.15	0.49 ± 0.01	0.46 ± 0.01

Table A3. Average S_{chain} of lipids in the equimolar asymmetric bilayers in the presence and absence of cholesterol. The "i" and "o" subscripts represent the inner and outer leaflets, respectively.

| | 0% Cholesterol | | 33% Cholesterol | | |
| | S_{Chain} | | S_{Chain} | | Cholesterol |
Lipid bilayer	Inner	Outer	Inner	Outer	Abundance
DPPCi/DPPCo/CHOL	0.40 ± 0.02	0.40 ± 0.02	0.69 ± 0.02	0.68 ± 0.02	Equal
16:0SMi/DPPCo/CHOL	0.40 ± 0.02	0.41 ± 0.02	0.69 ± 0.02	0.62 ± 0.02	Inner
DPPEi/DPPCo/CHOL	0.42 ± 0.02	0.42 ± 0.02	0.71 ± 0.02	0.69 ± 0.02	Inner
DPPSi/DPPCo/CHOL	0.38 ± 0.02	0.37 ± 0.02	0.69 ± 0.02	0.66 ± 0.02	Inner
DOPCi/DPPCo/CHOL	0.27 ± 0.02	0.32 ± 0.02	0.34 ± 0.02	0.64 ± 0.02	Outer
POPCi/DPPCo/CHOL	0.32 ± 0.02	0.35 ± 0.02	0.45 ± 0.02	0.67 ± 0.02	Outer
12:0PCi/DPPCo/CHOL	0.39 ± 0.02	0.40 ± 0.02	0.69 ± 0.02	0.69 ± 0.02	Outer
24:0PCi/DPPCo/CHOL	0.42 ± 0.02	0.40 ± 0.02	0.66 ± 0.02	0.65 ± 0.02	Inner
DOPEi/DPPCo/CHOL	0.28 ± 0.02	0.33 ± 0.02	0.35 ± 0.02	0.64 ± 0.02	Outer
POPEi/DPPCo/CHOL	0.34 ± 0.02	0.37 ± 0.02	0.48 ± 0.02	0.66 ± 0.02	Outer
DOPSi/DPPCo/CHOL	0.26 ± 0.02	0.30 ± 0.02	0.35 ± 0.02	0.60 ± 0.02	Outer
POPSi/DPPCo/CHOL	0.31 ± 0.02	0.33 ± 0.02	0.47 ± 0.02	0.61 ± 0.02	Outer

Table A4. Comparative data of lipid bilayers: we have studied symmetric bilayers, equimolar asymmetric bilayers (denoted by **) and asymmetric bilayers with no build-in stress (denoted by *).

| | Initial distribution | | | | Final distribution | | | | Chol% | | area/ lipid | | S_{Chain} | |
| | # Lipids | | # Chol | | # Lipids | | # Chol | | ± 1% | | ± 0.01 nm^2 | | ± 0.02 | |
Lipid bilayers	in	out	in	out	in	out	in	out	inner	outer	inner	outer	inner	outer
DOPCi/DOPCo/CHOL	170	170	85	85	170	170	85	85	50	50	0.53	0.53	0.31	0.31
DPPCi/DPPCo//CHOL	170	170	85	85	170	170	87	83	51	49	0.45	0.45	0.70	0.68
DOPCi/DPPCo/CHOL (*)	192	240	96	120	192	240	102	114	47	53	0.54	0.45	0.28	0.69
DOPCi/DPPCo/CHOL (**)	171	171	84	84	170	170	62	106	37	63	0.53	0.45	0.34	0.64
POPCi/POPCo/CHOL	170	170	85	85	170	170	86	84	50	50	0.50	0.50	0.43	0.43
DPPCi/DPPCo//CHOL	170	170	85	85	170	170	87	83	51	49	0.45	0.45	0.70	0.68
POPCi/DPPCo/CHOL (*)	216	240	108	120	216	240	107	121	47	53	0.50	0.45	0.43	0.69
POPCi/DPPCo/CHOL (**)	170	170	85	85	170	170	69	101	41	59	0.50	0.44	0.45	0.67
DOPEi/DOPEo/CHOL	170	170	85	85	170	170	85	85	50	50	0.52	0.52	0.33	0.33
DPPCi/DPPCo//CHOL	170	170	85	85	170	170	87	83	51	49	0.45	0.45	0.70	0.68
DOPEi/DPPCo/CHOL (*)	192	240	96	120	197	235	101	115	47	53	0.52	0.45	0.32	0.69
DOPEi/DPPCo/CHOL (**)	171	171	84	84	171	171	64	104	38	62	0.53	0.45	0.35	0.64
POPEi/POPEo/CHOL	170	170	85	85	170	170	87	83	51	49	0.49	0.49	0.47	0.45
DPPCi/DPPCo//CHOL	170	170	85	85	170	170	87	83	51	49	0.45	0.45	0.70	0.68
POPEi/DPPCo/CHOL (*)	216	240	108	120	216	240	111	117	49	51	0.49	0.45	0.46	0.69
POPEi/DPPCo/CHOL (**)	170	170	85	85	170	170	73	97	43	57	0.49	0.45	0.48	0.66
DPPEi/DPPEo/CHOL	170	170	85	85	170	170	82	88	48	52	0.45	0.44	0.70	0.73
DPPCi/DPPCo//CHOL	170	170	85	85	170	170	87	83	51	49	0.45	0.45	0.70	0.68
DPPEi/DPPCo/CHOL (*)	240	240	120	120	240	240	124	116	52	48	0.44	0.45	0.72	0.69
DPPEi/DPPCo/CHOL (**)	171	171	84	84	171	171	86	82	51	49	0.45	0.45	0.71	0.69
16:0SMi/16:0SMo/CHOL	170	170	85	85	171	169	87	83	51	49	0.44	0.45	0.69	0.66
DPPCi/DPPCo//CHOL	170	170	85	85	170	170	87	83	51	49	0.45	0.45	0.70	0.68
16:0SMi/DPPCo/CHOL (*)	240	240	120	120	240	240	129	111	54	46	0.44	0.46	0.68	0.65
16:0SMi/DPPCo/CHOL (**)	171	171	84	84	171	171	93	75	56	44	0.44	0.46	0.69	0.62
12:0PCi/12:0PCo/CHOL	170	170	85	85	175	165	79	91	47	53	0.45	0.45	0.70	0.69
DPPCi/DPPCo//CHOL	170	170	85	85	170	170	87	83	51	49	0.45	0.45	0.70	0.68
12:0PCi/DPPCo/CHOL (*)	240	240	120	120	240	240	117	123	49	51	0.46	0.45	0.68	0.69
12:0PCi/DPPCo/CHOL (**)	171	171	84	84	171	171	83	85	49	51	0.45	0.45	0.69	0.69

Table A5. Partial molar area of cholesterol and two-chain lipids in symmetric DOPC/CHOL and DPPC/CHOL bilayers. The $a(x)$ for the DOPC/CHOL bilayer varies linearly with x implying that the $a_{DOPC}(x)$ and $a_{chol}(x)$ are constants. On the other hand, the $a(x)$ for the DPPC/CHOL bilayer shows some curvature implying that the partial molar areas are not constants for this bilayer.

Lipid Bilayer	Cholesterol mole fraction (x)	System area (nm²)	$a(x), nm^2$	$a_{chol}(x), nm^2$	$a_{lipid}(x), nm^2$	S_{Chain}
			± 0.01	± 0.01	± 0.01	± 0.02
DOPC/CHOL	0	177.68 ± 1.59	0.69		0.69	0.23
	10	164.25 ± 1.60	0.64	0.21	0.69	0.25
	20	151.16 ± 1.42	0.59	0.21	0.69	0.27
	30	139.03 ± 1.33	0.55	0.21	0.69	0.30
	33	134.92 ± 1.35	0.53	0.21	0.69	0.31
	40	127.09 ± 1.21	0.50	0.21	0.69	0.33
	50	115.96 ± 1.12	0.45	0.21	0.69	0.37
DPPC/CHOL	0	157.78 ± 1.52	0.62		0.62	0.40
	10	143.32 ± 1.49	0.56	0.07	0.61	0.47
	20	129.40 ± 1.24	0.51	0.13	0.60	0.56
	30	118.09 ± 0.93	0.46	0.18	0.58	0.66
	33	114.72 ± 0.85	0.45	0.20	0.58	0.69
	40	109.50 ± 0.67	0.43	0.23	0.56	0.72
	50	101.85 ± 0.42	0.40	0.27	0.53	0.79

Figure A1. A) Composition of the asymmetric lipid bilayer studied in this work. **B)** A snapshot of equilibrium configuration of the asymmetric lipid bilayer at 37°C. The top-view shown in the Figure S1B corresponds to the inner-leaflet while the bottom view (not shown) corresponds to the outer-leaflet. Ions and water molecules are omitted for the sake of clarity. The colors of headgroups shown in Figure S1B match the colors shown in the pie-chart of Figure S1A. Lipid chains are shown in white. **C)** The structure of lipids in the CG Martini force field representation. The second bead of the headgroups for all phospholipids is phosphate (PO4). The glycerol backbones (GL1 and Gl2) are colored light silver while the green beads show the sphingosine backbone (AM1 and AM2). The beads with the unsaturation are colored in purple. Bead types are as follows: (choline NC3: Q0, ethanolamine NH3: Qd, serine CN0: P5, phosphate PO4: Qa, glycerol GL1: Na, glycerol GL2: Na, amide AM1: P1, amide AM2: P5, and cholesterol-hydroxyl ROH: SP1). GL1, GL2, and AM2 contain a carbonyl oxygen group while AM1 contains a hydroxyl group. The definition of Q0, Qd, Qa. Na, P1, P5, and SP1 are presented in the methodology section while the full explanation is available in the original paper.[44]

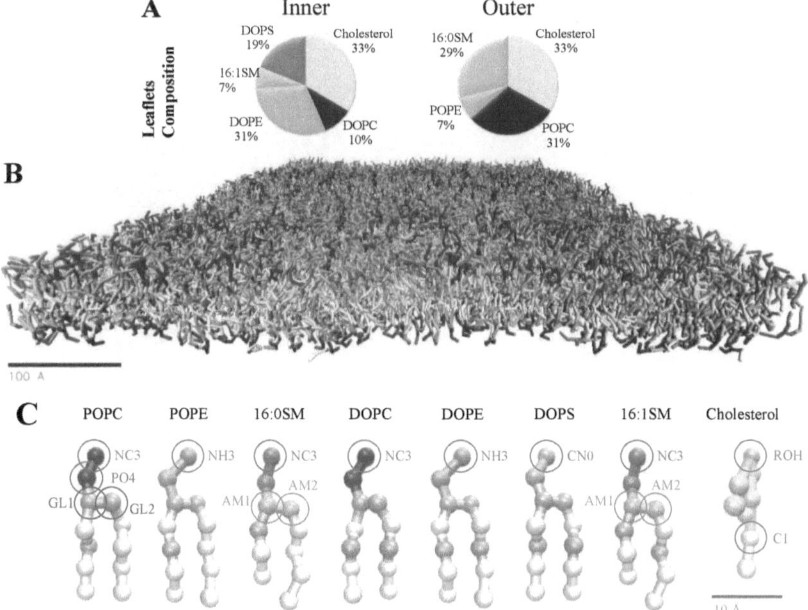

Figure A2. A) A snapshot of the equilibrated configuration of the DPPEi/DPPCo/CHOL system at 37°C. **B)** Structure of lipids with CG Martini force field representation to understand the effects of the backbone (16:0SM), headgroup type (DPPE and DPPS), acyl chain saturation (DOPC and POPC), and acyl chain length (12:0PC and 24:0PC). The second bead of the headgroups of all phospholipids is phosphate (PO4). The glycerol backbones (GL1 and Gl2) are colored light silver while the green beads show the sphingosine backbone (AM1 and AM2). The beads with unsaturation are colored in purple. Bead types are as follows: (choline NC3: Q0, ethanolamine NH3: Qd, serine CN0: P5, phosphate PO4: Qa, glycerol GL1: Na, glycerol GL2: Na, amide AM1: P1, amide AM2: P5, and cholesterol-hydroxyl ROH: SP1). According to the molecular structure, GL1, GL2, and AM2 contain a carbonyl oxygen group while AM1 contains a hydroxyl group. The definition of Q0, Qd, Qa. Na, P1, P5, and SP1 are presented in the methodology section while the full explanation is available in the original paper.[44]

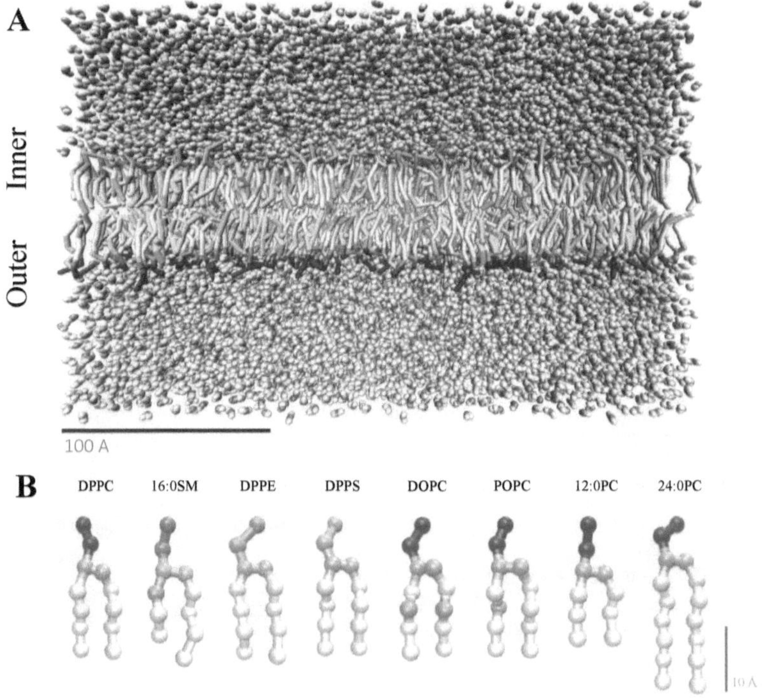

Figure A3. Density profile of phospholipids in the asymmetric system after 10 μs simulation. Negative values of z correspond to the outer leaflet while the positive values represent the inner leaflet.

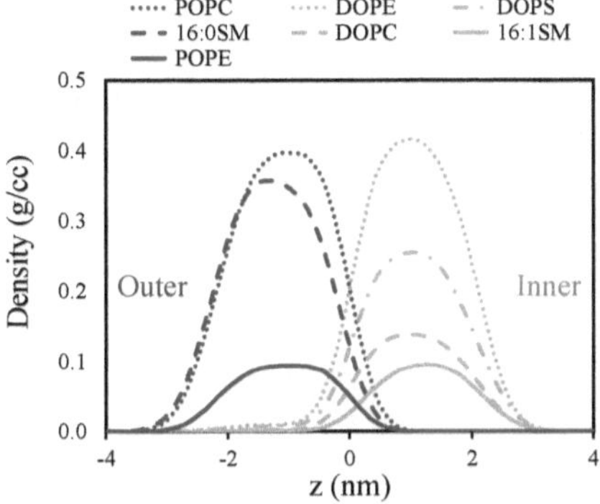

Appendix B: Correspond to the Materials Presented in Chapter 4

Table B1. Simulation parameters for all CG simulations in the present work using Dry and Wet martini force fields (see footnote for *abbreviations).

Simulation parameters	Study-I		Study-II
	Dry Martini	Wet Martini	Dry Martini
cutoff scheme	Verlet	Verlet	Verlet
Coulomb scheme	RF	RF	RF
cutoff, nm	1.2	1.1	1.2
dielectric constant (ε_r)	–	2.5	–
van-der-Waals scheme	cutoff	cutoff	cutoff
cutoff, nm	1.2	1.1	1.2
thermostat	VR	VR	VR
time constant (τ_T), ps	1.0	1.0	1.0
barostat	PR	PR	PR
time constant (τ_p), ps	4	12	4
reference p, bar	1.0	1.0	1.0
minimization algorithm	SD	SD	SD
number of steps	5×10^4	5×10^4	5×10^4
Pre-equilibration integrator	LF	LF	LF
time step (Δt), ps	10	10	10
number of steps	5×10^6	5×10^6	5×10^6
total time at each T, ns	50	50	50
production integrator	LF	LF	LF
time step (Δt), ps	5	20	5
number of steps	2×10^8	5×10^7	2×10^8
total time at each T, μs	1	1	1

*abbreviations: RF = reaction field; VR: velocity rescale developed by Bussi et al.[47]; PR = Parrinello-Rahman; SD = steepest decent; LF = leap frog.

Table B2. Comparison of Dry and Wet Martini force field to obtain area per lipid (apl) and membrane thickness (D_{P-P}) of pure DOPC and 16:0 SM lipid bilayers.

Lipid bilayer	T (K)	apl (nm^2)		D_{P-P} (Å)	
		Dry Martini	Wet Martini	Dry Martini	Wet Martini
Pure DOPC	300	0.67 ± 0.01	0.68 ± 0.01	37.12 ± 0.32	38.17 ± 0.24
	310	0.68 ± 0.01	0.69 ± 0.01	36.83 ± 0.31	37.92 ± 0.25
Pure 16:0SM	300	0.60 ± 0.01	0.59 ± 0.01	37.26 ± 0.30	38.87 ± 0.26
	320	0.63 ± 0.01	0.62 ± 0.01	36.43 ± 0.30	38.23 ± 0.27
	330	0.64 ± 0.01	0.63 ± 0.01	35.96 ± 0.32	37.93 ± 0.27

Figure B1. A) A snapshot of equilibrium configuration of the symmetric DOPC/16:0SM/CHOL lipid bilayer at 290 K. Ions and water molecules are omitted for the sake of clarity. Lipid chains are shown in white. **B)** The structure of lipids in the CG Martini force field representation. The second bead of the headgroups for all phospholipids is phosphate (PO4). The glycerol backbones (GL1 and Gl2) are colored light silver while the top two green beads in sphingomyelin molecules are the sphingosine backbone (AM1 and AM2). The beads with the unsaturation are colored in purple. Bead types are as follows: (choline NC3: Q0, phosphate PO4: Qa, glycerol GL1: Na, glycerol GL2: Na, amide AM1: P1, amide AM2: P5, and cholesterol-hydroxyl ROH: SP1). The full explanation of Martini bead types Q0, Qd, Qa. Na, P1, P5, and SP1 is available in the original paper.[155]

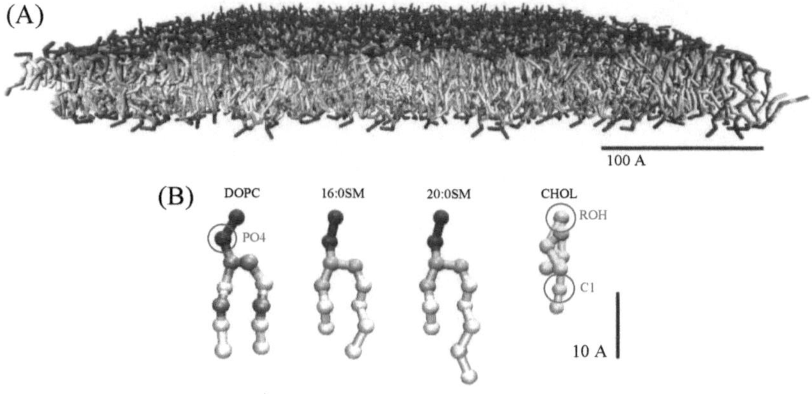

Figure B2. Comparison of Dry and Wet Martini force fields in investigation of structural properties of DOPC/CHOL lipid bilayer at temperature range of 350 K to 290 K, and cholesterol concentration of 30% mole-ratio. **A)** Area per lipid, *apl*; **B)** Bilayer thickness, D_{P-P}; **C)** Chain order parameter, S_{chain}; **D)** 2D RDFs of C1B-C1B beads of two-chain lipids for simulations performed with Dry Martini force field; and **E)** 2D RDFs of C1B-C1B beads of two-chain lipids for simulations performed with Wet Martini force field. Successive $g_{2D}(r)$ have been shifted vertically by 0.4 units for the sake of clarity. Error bars in all figures were estimated as standard error of the mean when the last 0.5 µs of each system were split into three equally sized blocks and analyzed separately.

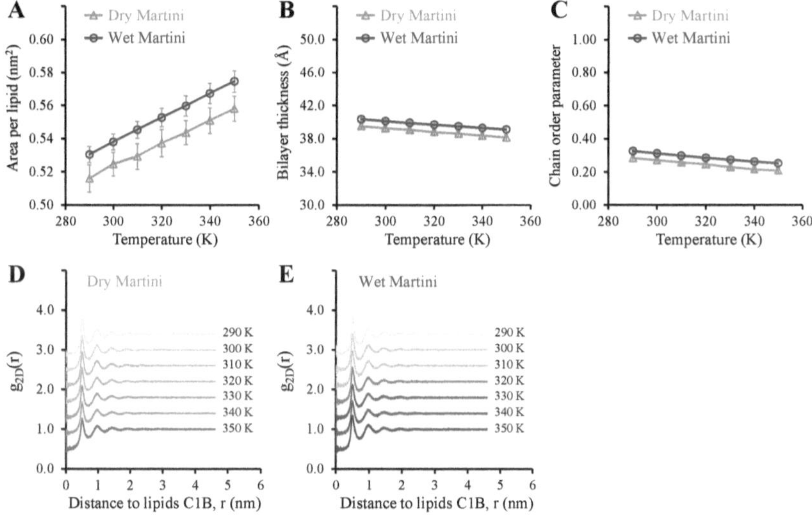

Figure B3. Comparison of Dry and Wet Martini force fields in investigation of structural properties of 16:0SM/CHOL lipid bilayer at temperature range of 350 K to 290 K, and cholesterol concentration of 30% mole-ratio. **A)** Area per lipid, *apl*; **B)** Bilayer thickness, D_{P-P}; **C)** Chain order parameter, S_{chain}; **D)** 2D RDFs of C1B-C1B beads of two-chain lipids for simulations performed with Dry Martini force field; and **E)** 2D RDFs of C1B-C1B beads of two-chain lipids for simulations performed with Wet Martini force field. Successive $g_{2D}(r)$ have been shifted vertically by 0.4 units for the sake of clarity. Error bars in all figures were estimated as standard error of the mean when the last 0.5 µs of each system were split into three equally sized blocks and analyzed separately.

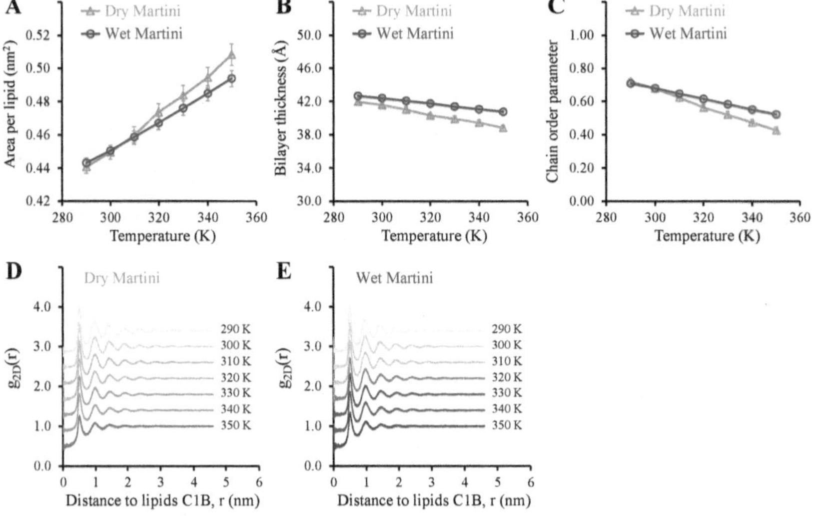

Figure B4. Phase transition temperature, Tm, calculation by finding derivation of **A)** Area per lipid, *apl*, **B)** Bilayer thickness, D_{P-P}, **B)** Chain order parameter, S_{chain}; and locate the relative maximum for **1)** pure DOPC, **2)** pure 20:0SM, and **3)** 1:1 DOPC/20:0SM lipid bilayers.

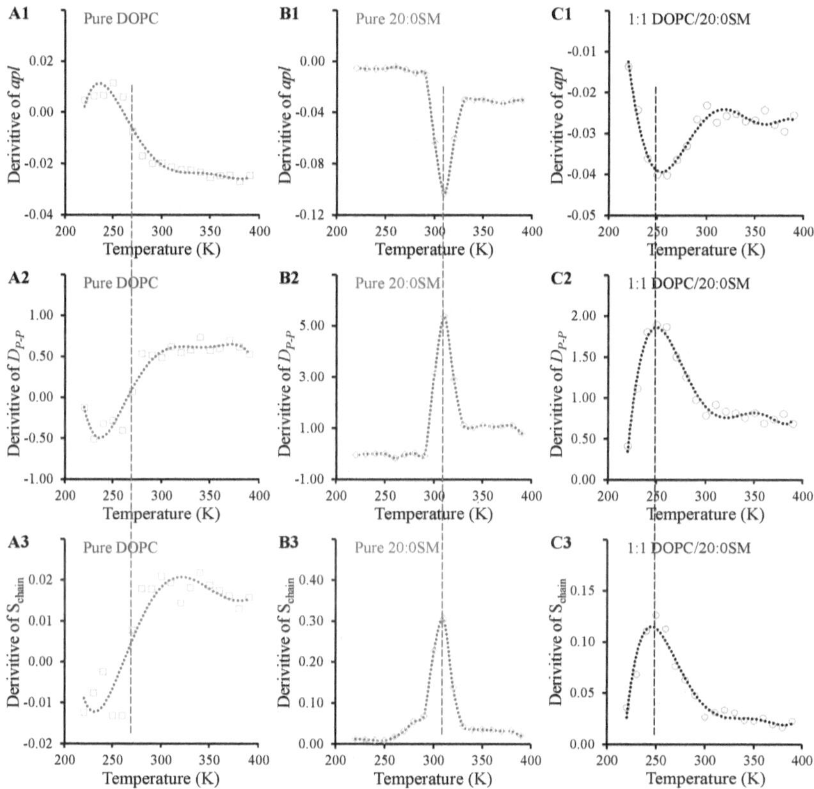

Figure B5. Structural properties of **A)** DOPC/CHOL and **B)** DOPC/20:0SM/CHOL bilayers as a function of temperature and cholesterol concentration. **1)** Area per lipid, *apl*; **2)** Bilayer thickness, D_{P-P}; and **3)** chain order parameter, S_{chain}. Error bars in all figures were estimated as standard error of the mean when the last 0.5 μs of each system were split into three equally sized blocks and analyzed separately.

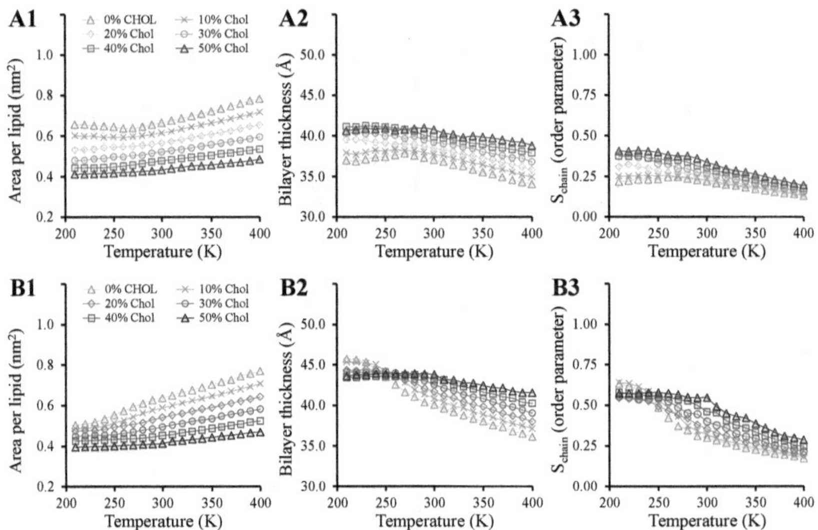

Figure B6. Phase transition temperature, *Tm*, calculation by finding derivation of chain order parameter, S_{chain}, and locate the relative maximum for **A)** DOPC/CHOL, **B)** 20:0SM/CHOL, and **C)** DOPC/20:0SM /CHOL lipid bilayers with cholesterol mole-ratios of **1)** 10%, **2)** 20%, **3)** 30%, **4)** 40%, and **5)** 50%.

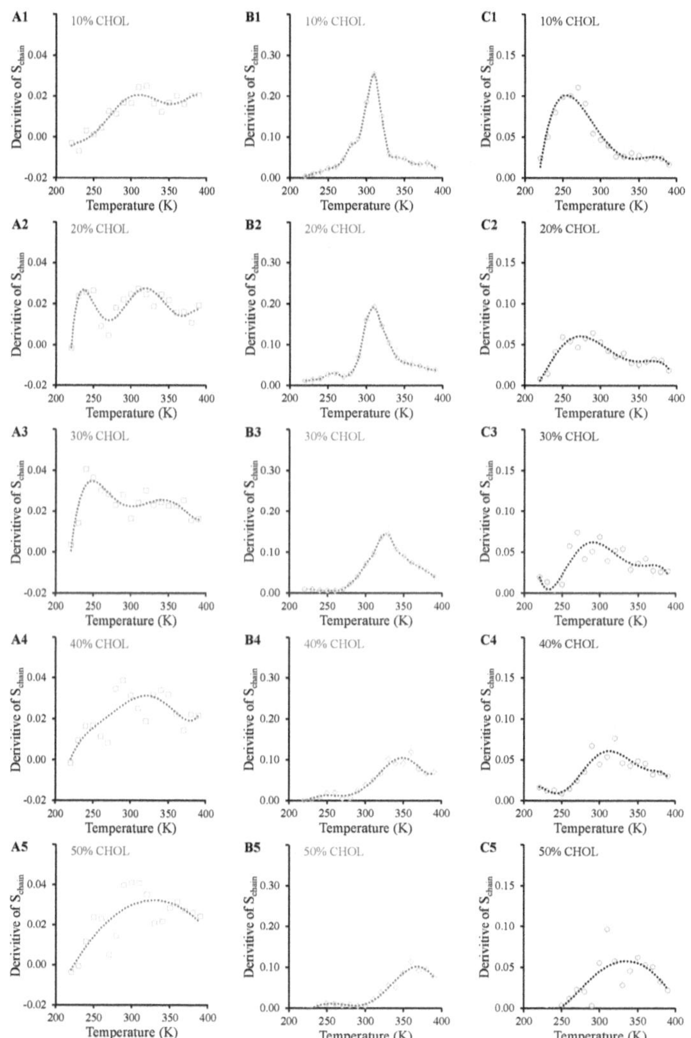

Figure B7. Ensemble-averaged angle of cholesterol molecule with the bilayer normal as a function of the location of the hydroxyl headgroup of cholesterol in the three lipid bilayers of **A)** DOPC/CHOL, **B)** 20:0SM/CHOL, and **C)** DOPC/20:0SM/CHOL with cholesterol mole-ratios of **1)** 10%, **2)** 20%, **3)** 30%, **4)** 40%, and **5)** 50%. Shaded area represents the midplane of the bilayer (13Å, 14Å, and 15Å from the center of the bilayers in figures A, B, and C, respectively). Error bars in all figures were estimated as standard error of the mean when the last 0.5 μs of each system were split into three equally sized blocks and analyzed separately.

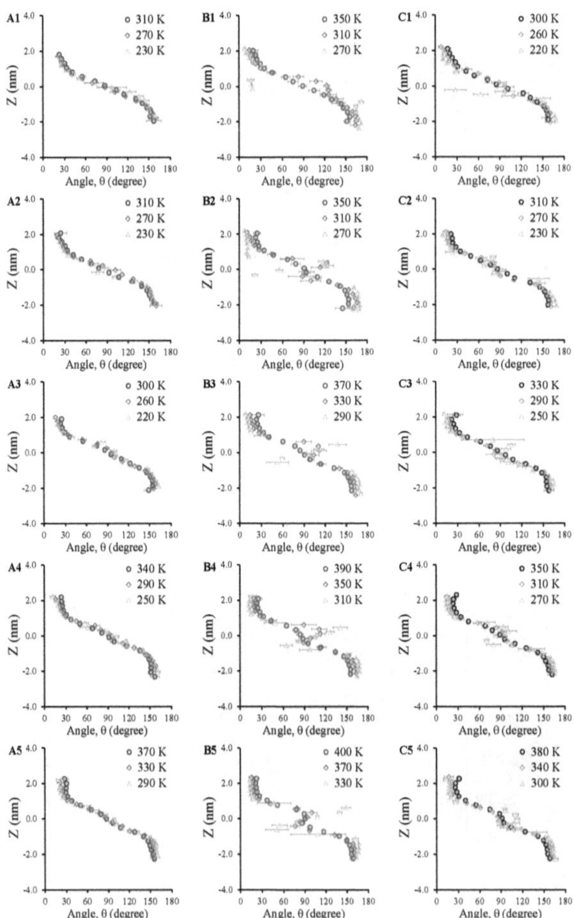

Figure B8. Cross temperature, *Tcross*, calculation based on the fraction of cholesterol molecules at different regions of the bilayers for **A)** DOPC/CHOL, **B)** 20:0SM/CHOL, and **C)** DOPC/20:0SM/CHOL systems with cholesterol mole-ratios of **1)** 10%, **2)** 20%, **3)** 30%, **4)** 40%, and **5)** 50%. Error bars in all figures were estimated as standard error of the mean when the last 0.5 μs of each system were split into three equally sized blocks and analyzed separately. Table 4 (in the text of main manuscript) summarizes all the *Tcross* values numerically.

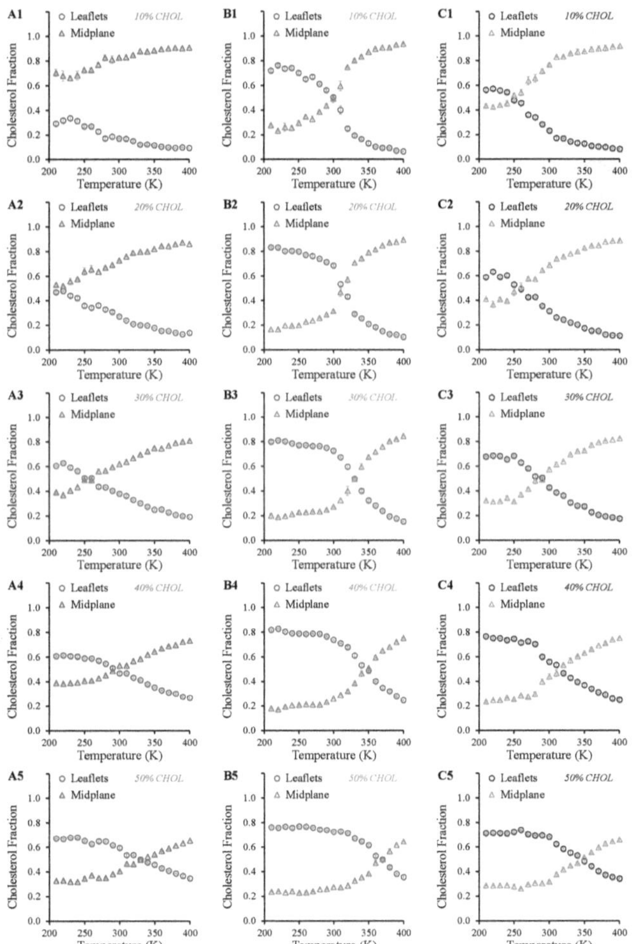

Appendix C: Correspond to the Materials Presented in Chapter 5

Table C1. Melting temperature (*Tm*) dataset. All Tm values, are from the Lipid Thermotropic Phase Transition Database (LIPIDAT) – NIST Standard Reference Database 34 [111,112], unless otherwise noted. The lipids that were randomly selected for the training dataset are denoted by symbol *. Features (columns) definition: *T1L* (tail 1 length, varied between 3 and 24), *T2L* (tail 2 length, varied between 3 and 24), *MW* (molecular weight in g/mole), *BT* (backbone type, glycerol = 1, sphingosine = 2), *HS* (head size, choline = 104 g/mole, ethanolamine = 61 g/mole, glycerol = 92 g/mole, serine = 105 g/mole, hydroxyl = 17 g/mole), *ChT* (acyl chain type, not mixed acyl chain = 0, mixed acyl chain = 1), *ST* (saturation type, saturated = 0, unsaturated-cis = 1, unsaturated-trans = 2), *NU* (number of unsaturated carbons, varied between 0 and 12), *HCh* (head charge, zwitterionic = 0, anionic = -1), *ID1* (first unsaturated carbon ID number on tail 1), *ID2* (first unsaturated carbon ID number on tail 2), and *Tm* (melting temperature in degree C).

Lipid Name (Abbreviation)	T1L	T2L	MW	BT	HS	ChT	ST	NU	HCh	ID1	ID2	TM
12:0 PC (DLPC)	12	12	622	1	104	0	0	0	0	0	0	-2
13:0 PC	13	13	650	1	104	0	0	0	0	0	0	14
14:0 PC (DMPC)	14	14	678	1	104	0	0	0	0	0	0	24
15:0 PC	15	15	706	1	104	0	0	0	0	0	0	35
16:0 PC (DPPC) *	16	16	734	1	104	0	0	0	0	0	0	41
17:0 PC *	17	17	762	1	104	0	0	0	0	0	0	50
18:0 PC (DSPC)	18	18	790	1	104	0	0	0	0	0	0	55
19:0 PC	19	19	818	1	104	0	0	0	0	0	0	62
20:0 PC	20	20	846	1	104	0	0	0	0	0	0	66
21:0 PC *	21	21	874	1	104	0	0	0	0	0	0	71
22:0 PC	22	22	902	1	104	0	0	0	0	0	0	75
23:0 PC	23	23	930	1	104	0	0	0	0	0	0	79.5
24:0 PC	24	24	958	1	104	0	0	0	0	0	0	80.3
16:1 (Δ9-Cis) PC	16	16	730	1	104	0	1	2	0	9	9	-36
18:1 (Δ6-Cis) PC	18	18	786	1	104	0	1	2	0	6	6	1
18:1 (Δ9-Cis) PC (DOPC) *	18	18	786	1	104	0	1	2	0	9	9	-17
18:1 (Δ9-Trans) PC	18	18	786	1	104	0	2	2	0	9	9	12
18:2 PC (DLiPC)	18	18	782	1	104	0	1	4	0	9	9	-57
18:3 PC *	18	18	778	1	104	0	1	6	0	9	9	-60
20:4 PC (DAPC)	20	20	830	1	104	0	1	8	0	5	5	-69
22:1 PC (DEPC)	22	22	898	1	104	0	1	2	0	13	13	13
14:0 - 16:0 PC (MPPC)	14	16	706	1	104	1	0	0	0	0	0	35
14:0 - 18:0 PC (MSPC) *	14	18	734	1	104	1	0	0	0	0	0	40
16:0 - 14:0 PC (PMPC) *	16	14	706	1	104	1	0	0	0	0	0	27
16:0 - 18:0 PC (PSPC)	16	18	762	1	104	1	0	0	0	0	0	49
16:0 - 18:1 PC (POPC)	16	18	760	1	104	1	1	1	0	0	9	-2

16:0 - 22:6 PC	16	22	806	1	104	1	1	6	0	0	4	-27
18:0 - 14:0 PC (SMPC)	18	14	734	1	104	1	0	0	0	0	0	30
18:0 - 16:0 PC (SPPC)	18	16	762	1	104	1	0	0	0	0	0	44
18:0 - 18:1 PC (SOPC)	18	18	788	1	104	1	1	1	0	0	9	6
18:0 - 22:6 PC (SDPC) *	18	22	834	1	104	1	1	6	0	0	4	-6 [a]
18:1 - 16:0 PC (OPPC)	18	16	760	1	104	1	1	1	0	9	0	-9
18:1 - 18:0 PC (OSPC)	18	18	788	1	104	1	1	1	0	9	0	9
12:0 PE (DLPE)	12	12	580	1	61	0	0	0	0	0	0	29
14:0 PE (DMPE)	14	14	636	1	61	0	0	0	0	0	0	50
16:0 PE (DPPE)	16	16	692	1	61	0	0	0	0	0	0	63
18:0 PE (DSPE)	18	18	748	1	61	0	0	0	0	0	0	74
20:0 PE	20	20	804	1	61	0	0	0	0	0	0	83
18:1 (Δ9-Cis) PE (DOPE)	18	18	744	1	61	0	1	2	0	9	9	-16
18:1 (Δ9-Trans) PE	18	18	744	1	61	0	2	2	0	9	9	38
18:2 PE (DLiPE)	18	18	740	1	61	0	1	4	0	9	9	-40
16:0 - 18:1 PE (POPE)	16	18	718	1	61	1	1	1	0	0	9	25
12:0 PG (DLPG)	12	12	633	1	92	0	0	0	-1	0	0	-3
14:0 PG (DMPG)	14	14	689	1	92	0	0	0	-1	0	0	23
16:0 PG (DPPG)	16	16	745	1	92	0	0	0	-1	0	0	41
18:0 PG (DSPG)	18	18	801	1	92	0	0	0	-1	0	0	55
18:1 (Δ9-Cis) PG (DOPG)	18	18	797	1	92	0	1	2	-1	9	9	-18
16:0 - 18:1 PG (POPG)	16	18	771	1	92	1	1	1	-1	0	9	-2
14:0 PS (DMPS)	14	14	702	1	105	0	0	0	-1	0	0	35
16:0 PS (DPPS)	16	16	758	1	105	0	0	0	-1	0	0	54
18:0 PS (DSPS)	18	18	814	1	105	0	0	0	-1	0	0	68
18:1 (Δ9-Cis) PS (DOPS)	18	18	810	1	105	0	1	2	-1	9	9	-11
16:0 - 18:1 PS (POPS)	16	18	784	1	105	1	1	1	-1	0	9	14
12:0 PA (DLPA)	12	12	559	1	17	0	0	0	-1	0	0	31
14:0 PA (DMPA)	14	14	615	1	17	0	0	0	-1	0	0	52
16:0 PA (DPPA)	16	16	671	1	17	0	0	0	-1	0	0	65
18:1 (Δ9-Cis) PA (DOPA)	18	18	723	1	17	0	1	2	-1	9	9	-4
16:0 - 18:1 PA (POPA)	16	18	697	1	17	1	1	1	-1	0	9	28
14:0 SM (d18:1/14:0)	18	14	675	2	104	1	0	0	0	0	0	25.4
16:0 SM (d18:1/16:0) (PSM)	18	16	703	2	104	1	0	0	0	0	0	40.5 [b]
18:0 SM (d18:1/18:0)	18	18	731	2	104	1	0	0	0	0	0	44.7
24:0 SM (d18:1/24:0)	18	24	815	2	104	1	0	0	0	0	0	47.1
Egg SM	18	16.57	711	2	104	1	0	0	0	0	0	38 [c]
Brain SM	18	20.08	760	2	104	1	0	0	0	0	0	40 [d]
Milk SM *	18	21.85	785	2	104	1	0	0	0	0	0	34.3 [e]

a from [156], *b* from[88,95], *c* from[157], *d* from[79], and *e* from[158]

Table C2. An abbreviated version of phase diagram training dataset. Features (columns) definition: *T1L1* (lipid1 tail1 length), *T2L1* (lipid1 tail2 length), *ST1* (lipid 1 saturation type), *NU1* (lipid 1 number of unsaturated carbons), *MW1* (lipid1 molecular weight), *TM1* (lipid1 melting temperature), *MF1* (lipid1 mole fraction), *T1L2* (lipid2 tail1 length), *T2L2* (lipid2 tail2 length), *ST2* (lipid 2 saturation type), *NU2* (lipid 2 number of unsaturated carbons), *MW2* (lipid2 molecular weight), *TM2* (lipid2 melting temperature), *MF2* (lipid2 mole fraction), *T* (temperature of the system), and *Phase* (response variable; phase of the mixture). Number codes: 1: *Ld*, 2: *Lo*, 3: *Lβ*, 4: *Ld+Lo*, 5: *Ld+Lβ*, 6: *Lo+Lβ*, 7: *Ld+Lo+Lβ*, and 8: *Crystals+Lo*.

Sample #	T1L1	T2L1	ST1	NU1	MW1	TM1	MF1	T1L2	T2L2	ST2	NU2	MW2	MT2	MF2	T	Phase
1	12	12	0	0	622	-2	0	16	16	0	0	734	41	0	24	8
2	12	12	0	0	622	-2	1	16	16	0	0	734	41	0	24	8
3	12	12	0	0	622	-2	0	16	16	0	0	734	41	1	24	8
4	12	12	0	0	622	-2	2	16	16	0	0	734	41	0	24	8
...																
...																
9360	16	18	1	1	760	-2	69	18	18	0	0	790	55	22	23	7
9361	16	18	1	1	760	-2	68	18	18	0	0	790	55	23	23	5
9362	16	18	1	1	760	-2	67	18	18	0	0	790	55	24	23	5
...																
...																
18154	18	18	1	2	786	-17	0	16	18	1	1	760	-2	72	23	1
18155	18	18	1	2	786	-17	73	16	18	1	1	760	-2	0	23	2
18256	18	18	1	2	786	-17	72	16	18	1	1	760	-2	1	23	2
...																
...																
36055	18	18	1	2	786	-17	2	16	16	0	0	734	41	98	18	3
36056	18	18	1	2	786	-17	1	16	16	0	0	734	41	99	18	3
36057	18	18	1	2	786	-17	0	16	16	0	0	734	41	100	18	3

Table C3. The L_{25} orthogonal array for the Taguchi method.

Experiment	Factor level				
	OS	US	NN	NV/Cost	NT/Gamma
1	1	1	1	1	1
2	1	2	2	2	2
3	1	3	3	3	3
4	1	4	4	4	4
5	1	5	5	5	5
6	2	1	2	3	4
7	2	2	3	4	5
8	2	3	4	5	1
9	2	4	5	1	2
10	2	5	1	2	3
11	3	1	3	5	2
12	3	2	4	1	3
13	3	3	5	2	4
14	3	4	1	3	5
15	3	5	2	4	1
16	4	1	4	2	5
17	4	2	5	3	1
18	4	3	1	4	2
19	4	4	2	5	3
20	4	5	3	1	4
21	5	1	5	4	3
22	5	2	1	5	4
23	5	3	2	1	5
24	5	4	3	2	1
25	5	5	4	3	2

Table C4. Tm values, reported by the ANN model, for the lipids not experimentally evaluated in the literature. Features (columns) definition: *T1L* (tail 1 length, varied between 3 and 24), *T2L* (tail 2 length, varied between 3 and 24), *MW* (molecular weight in g/mole), *BT* (backbone type, glycerol = 1, sphingosine = 2), *HS* (head size, choline = 104 g/mole, ethanolamine = 61 g/mole, glycerol = 92 g/mole, serine = 105 g/mole, hydroxyl = 17 g/mole), *ChT* (acyl chain type, not mixed acyl chain = 0, mixed acyl chain = 1), *ST* (saturation type, saturated = 0, unsaturated-cis = 1, unsaturated-trans = 2), *NU* (number of unsaturated carbons, varied between 0 and 12), *HCh* (head charge, zwitterionic = 0, anionic = -1), *ID1* (first unsaturated carbon ID number on tail 1), *ID2* (first unsaturated carbon ID number on tail 2), and *Tm* (melting temperature in °C). Note that the *Tm* for PE lipids has not been predicted using the model given the complex polymorphic phase behavior of this lipids and its dependence on water content.[119,120]

Lipid Name (Abbreviation)	T1L	T2L	MW	BT	HS	ChT	ST	NU	HCh	ID1	ID2	Tm
3:0 PC	3	3	369	1	104	0	0	0	0	0	0	-5.5
4:0 PC	4	4	397	1	104	0	0	0	0	0	0	-5.5
5:0 PC	5	5	425	1	104	0	0	0	0	0	0	-5.5
6:0 PC (DHxPC)	6	6	454	1	104	0	0	0	0	0	0	-5.5
7:0 PC (DHpPC)	7	7	482	1	104	0	0	0	0	0	0	-5.4
8:0 PC	8	8	510	1	104	0	0	0	0	0	0	-5.4
9:0 PC	9	9	538	1	104	0	0	0	0	0	0	-5.3
10:0 PC	10	10	566	1	104	0	0	0	0	0	0	-5.0
11:0 PC	11	11	594	1	104	0	0	0	0	0	0	-4.5
14:1 (Δ9-Cis) PC	14	14	674	1	104	0	1	2	0	9	9	-46.9
14:1 (Δ9-Trans) PC	14	14	674	1	104	0	2	2	0	9	9	25.6
16:1 (Δ9-Trans) PC	16	16	730	1	104	0	2	2	0	9	9	31.5
20:1 PC	20	20	842	1	104	0	1	2	0	11	11	-18.8
24:1 PC	24	24	954	1	104	0	1	2	0	15	15	25.6
22:6 PC	22	22	878	1	104	0	1	12	0	4	4	-57.9
16:0 - 02:0 PC	16	2	538	1	104	1	0	0	0	0	0	-5.3
16:0 - 18:2 PC (PLiPC)	16	18	758	1	104	1	1	2	0	0	9	-6.8
16:0 - 20:4 PC (PAPC)	16	20	782	1	104	1	1	4	0	0	5	-6.8
18:0 - 18:2 PC (SLiPC)	18	18	786	1	104	1	1	2	0	0	9	-4.9
18:0 - 20:4 PC (SAPC)	18	20	810	1	104	1	1	4	0	0	5	-5.4
18:1 - 14:0 PC (OMPC)	18	14	732	1	104	1	1	1	0	9	0	-24.6
6:0 PG	6	6	464	1	92	0	0	0	-1	0	0	-5.4
8:0 PG	8	8	521	1	92	0	0	0	-1	0	0	-5.3
10:0 PG	10	10	577	1	92	0	0	0	-1	0	0	-4.3
15:0 PG	15	15	717	1	92	0	0	0	-1	0	0	35.6
17:0 PG	17	17	773	1	92	0	0	0	-1	0	0	55.6
18:1 (Δ9-Trans) PG	18	18	797	1	92	0	2	2	-1	9	9	59.7
18:2 PG (DLiPG)	18	18	793	1	92	0	1	4	-1	9	9	-45.9
18:3 PG	18	18	789	1	92	0	1	6	-1	9	9	-51.9

20:4 PG (DAPG)	20	20	841	1	92	0	1	8	-1	5	5	-48.6
22:6 PG	22	22	889	1	92	0	1	12	-1	4	4	-49.6
16:0 - 18:2 PG (PLiPG)	16	18	769	1	92	1	1	2	-1	0	9	-5.3
16:0 - 20:4 PG (PAPG)	16	20	793	1	92	1	1	4	-1	0	5	-4.8
16:0 - 22:6 PG	16	22	817	1	92	1	1	6	-1	0	4	-8.2
18:0 - 18:1 PG (SOPG)	18	18	799	1	92	1	1	1	-1	0	9	-3.3
18:0 - 18:2 PG (SLiPG)	18	18	797	1	92	1	1	2	-1	0	9	-5.4
18:0 - 20:4 PG (SAPG)	18	20	821	1	92	1	1	4	-1	0	5	-5.2
18:0 - 22:6 PG	18	22	845	1	92	1	1	6	-1	0	4	-9.7
6:0 PS	6	6	477	1	105	0	0	0	-1	0	0	-5.5
8:0 PS	8	8	534	1	105	0	0	0	-1	0	0	-5.2
10:0 PS	10	10	590	1	105	0	0	0	-1	0	0	-3.8
12:0 PS (DLPS)	12	12	646	1	105	0	0	0	-1	0	0	4.0
17:0 PS	17	17	786	1	105	0	0	0	-1	0	0	51.6
18:2 PS (DLiPS)	18	18	806	1	105	0	1	4	-1	9	9	-44.8
20:4 PS (DAPS)	20	20	854	1	105	0	1	8	-1	5	5	-47.3
22:6 PS	22	22	902	1	105	0	1	12	-1	4	4	-48.4
16:0 - 18:2 PS (PLiPS)	16	18	782	1	105	1	1	2	-1	0	9	-5.7
16:0 - 20:4 PS (PAPS)	16	20	806	1	105	1	1	4	-1	0	5	-5.2
16:0 - 22:6 PS	16	22	830	1	105	1	1	6	-1	0	4	-7.6
18:0 - 18:1 PS (SOPS)	18	18	812	1	105	1	1	1	-1	0	9	-4.5
18:0 - 18:2 PS (SLiPS)	18	18	810	1	105	1	1	2	-1	0	9	-5.8
18:0 - 20:4 PS (SAPS)	18	20	834	1	105	1	1	4	-1	0	5	-5.4
18:0 - 22:6 PS	18	22	858	1	105	1	1	6	-1	0	4	-8.9
6:0 PA	6	6	390	1	17	0	0	0	-1	0	0	-5.1
8:0 PA	8	8	446	1	17	0	0	0	-1	0	0	-4.3
10:0 PA	10	10	503	1	17	0	0	0	-1	0	0	-0.9
17:0 PA	17	17	699	1	17	0	0	0	-1	0	0	69.6
18:0 PA (DSPA)	18	18	727	1	17	0	0	0	-1	0	0	72.0
18:2 PA (DLiPA)	18	18	719	1	17	0	1	4	-1	9	9	-41.7
20:4 PA (DAPA)	20	20	767	1	17	0	1	8	-1	5	5	-55.3
22:6 PA	22	22	815	1	17	0	1	12	-1	4	4	-56.7
16:0 - 18:2 PA (PLiPA)	16	18	695	1	17	1	1	2	-1	0	9	10.0
16:0 - 20:4 PA (PAPA)	16	20	719	1	17	1	1	4	-1	0	5	2.9
16:0 - 22:6 PA	16	22	743	1	17	1	1	6	-1	0	4	-14.6
18:0 - 18:1 PA (SOPA)	18	18	725	1	17	1	1	1	-1	0	9	36.4
18:0 - 18:2 PA (SLiPA)	18	18	723	1	17	1	1	2	-1	0	9	14.2
18:0 - 20:4 PA (SAPA)	18	20	747	1	17	1	1	4	-1	0	5	4.1
18:0 - 22:6 PA	18	22	771	1	17	1	1	6	-1	0	4	-17.4
02:0 SM (d18:1/2:0)	18	2	507	2	104	1	0	0	0	0	0	-5.3
06:0 SM (d18:1/6:0)	18	6	563	2	104	1	0	0	0	0	0	-4.9
12:0 SM (d18:1/12:0)	18	12	647	2	104	1	0	0	0	0	0	8.9

16:1 SM (d18:1/16:1)	18	16	701	2	104	1	1	1	0	0	9	-10.9
17:0 SM (d18:1/17:0)	18	17	717	2	104	1	0	0	0	0	0	40.4
18:1 SM (d18:1/18:1)	18	18	729	2	104	1	1	1	0	0	9	-7.5

Figure C1. Ternary phase diagrams of different phospholipid/cholesterol mixtures. Number/Color codes: 1/Dark blue: *Ld*, 2/Light blue: *Lo*, 3/Red: *Lβ*, 4/Cyan: *Ld+Lo*, 5/Green: *Ld+Lβ*, 6/yellow: *Lo+Lβ*, 7/Purple: *Ld+Lo+Lβ*, and 8/Black: *Crystals+Lo*. **A)** DLPC/DPPC/CHOL at 24°C;[84] **B)** POPC/DSPC/CHOL at 23°C;[80] **C)** DOPC/DSPC/CHOL at 23°C;[80] **D)** DOPC/POPC/CHOL at 23°C;[80] **E)** DOPC/DPPC/CHOL at 28°C;[87] **F)** DOPC/DPPC/CHOL at 22°C;[87] **G)** DOPC/DPPC/CHOL at 18°C;[87] and **H)** SDPC/BSM/CHOL at 23°C.[79] The phase diagram in H was used as the testing set. Lipid abbreviations: DLPC: 1,2-dilauroyl-sn-glycero-3-phosphocholine, DPPC: 1,2-dipalmitoyl-sn-glycero-3-phosphocholine, POPC: 1-palmitoyl-2-oleoyl-glycero-3-phosphocholine, DSPC: 1,2-distearoyl-sn-glycero-3-phosphocholine, DOPC: 1,2-dioleoyl-sn-glycero-3-phosphocholine, SDPC: 1-stearoyl-2-docosahexaenoyl-sn-glycero-3-phosphocholine, BSM: Sphingomyelin (Brain).

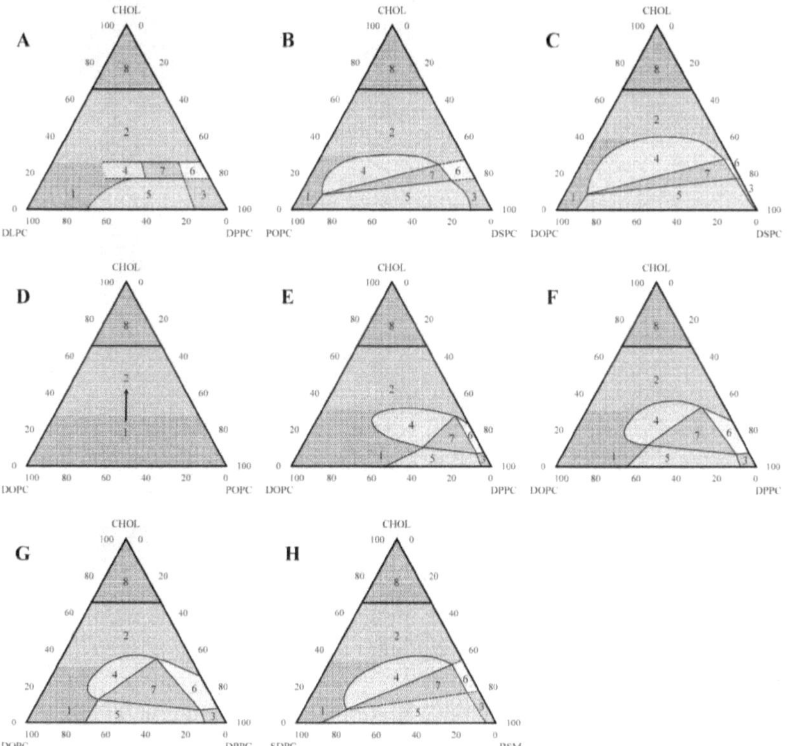

Figure C2. Demonstration of all 5,151 points for each phase diagram. MF1, MF2, and MF3 represents mole fraction of the lipid 1, lipid 2, and cholesterol, respectively. One point on the figure is shown as an example of all 5,151 points. **Point # 2511:** MF1/MF2/MF3 = 45/25/30.

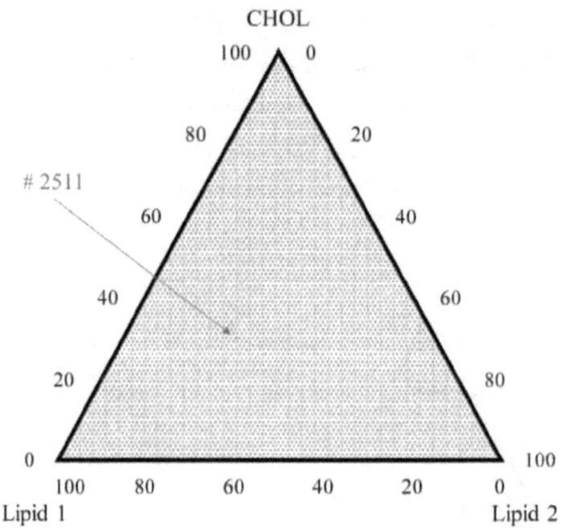

Figure C3. Main effects plots for signal-to-noise ratio for each level of the factors in **A)** the RF algorithm and **B)** the SVM algorithm (Larger is better).

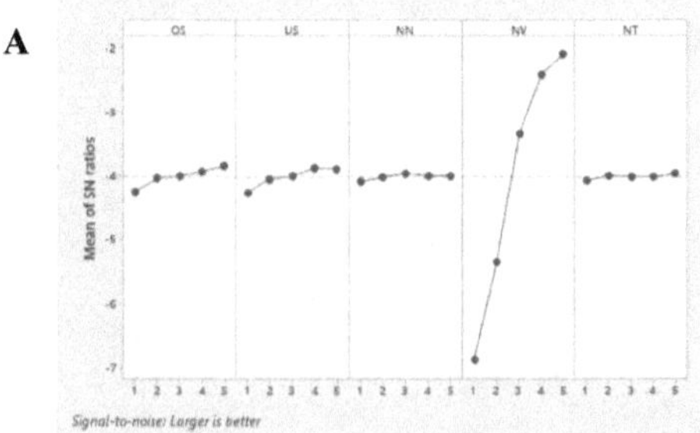

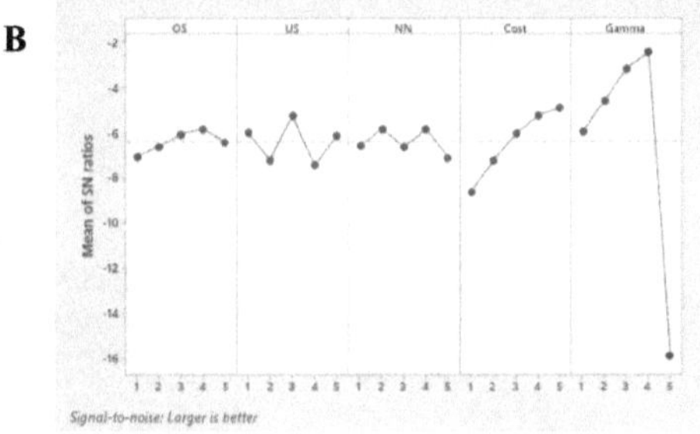

Figure C4. The phase diagrams for the SDPC/BSM/CHOL lipid mixture at 23 °C from **A)** Konyakhina et al.[79] **B)** the RF model, and **C)** the SVM model. Number/Color codes: 1/Dark blue: *Ld*, 2/Light blue: *Lo*, 3/Red: *Lβ*, 4/Cyan: *Ld+Lo*, 5/Green: *Ld+Lβ*, 6/yellow: *Lo+Lβ*, 7/Purple: *Ld+Lo+Lβ*, and 8/Black: *Crystals+Lo*. Lipid abbreviations: SDPC: *1-stearoyl-2-docosahexaenoyl-sn-glycero-3-phosphocholine*, BSM: *Sphingomyelin (Brain)*. Yellow dots show the phase boundaries based on which a triangle with straight sides was developed for the three-phase coexistence region.

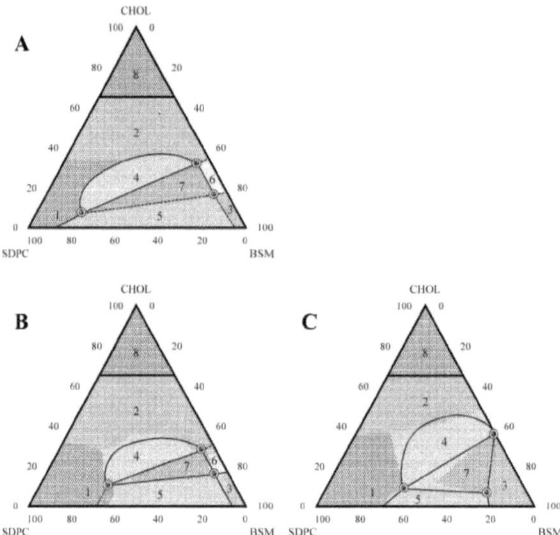

Figure C5. Detailed data points for the RF model predicted phase diagrams for DOPC/DPPC/CHOL system at **A)** 15°C and **B)** 37°C. Detailed data points for the RF model predicted phase diagrams for POPC/PSM/CHOL system at **C)** 23°C and **D)** 37°C. Number/Color codes: 1/Dark blue: *Ld*, 2/Light blue: *Lo*, 3/Red: *Lβ*, 4/Cyan: *Ld+Lo*, 5/Green: *Ld+Lβ*, 6/yellow: *Lo+Lβ*, 7/Purple: *Ld+Lo+Lβ*, and 8/Black: *Crystals+Lo*. Lipid abbreviations: DOPC: *1,2-dioleoyl-sn-glycero-3-phosphocholine*, DPPC: *1,2-dipalmitoyl-sn-glycero-3-phosphocholine*, POPC: *1-palmitoyl-2-oleoyl-glycero-3-phosphocholine*, PSM: *N-palmitoyl-D-erythro-sphingosylphosphorylcholine.* Yellow dots show the phase boundaries based on which a triangle with straight sides was developed for the three-phase coexistence region.

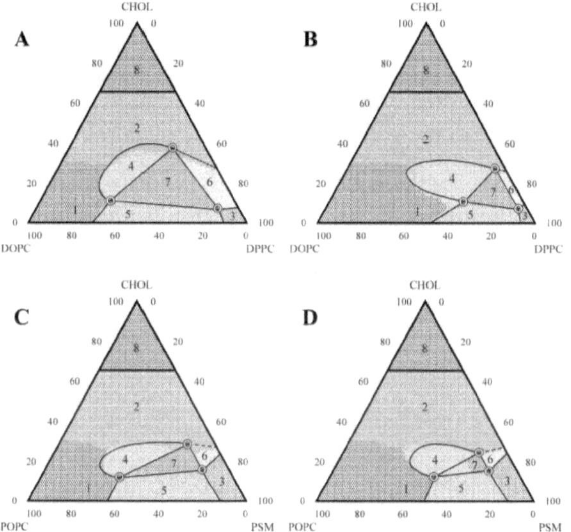

Figure C6. Detailed data points for the reported and predicted phase diagram for **A)** DOPC/DPPC/CHOL from [87] and **B)** POPC/DPPC/CHOL predicted by the RF model. Both diagrams are generated at a temperature of 22°C. Number/Color codes: 1/Dark blue: *Ld*, 2/Light blue: *Lo*, 3/Red: *Lβ*, 4/Cyan: *Ld+Lo*, 5/Green: *Ld+Lβ*, 6/yellow: *Lo+Lβ*, 7/Purple: *Ld+Lo+Lβ*, and 8/Black: *Crystals+Lo*. Lipid abbreviations: DOPC: *1,2-dioleoyl-sn-glycero-3-phosphocholine*, POPC: *1-palmitoyl-2-oleoyl-glycero-3-phosphocholine*, DPPC: *1,2-dipalmitoyl-sn-glycero-3-phosphocholine*. Yellow dots show the phase boundaries based on which a triangle with straight sides was developed for the three-phase coexistence region.

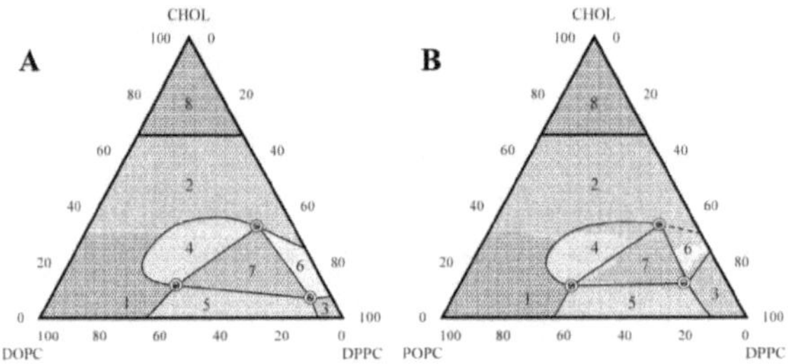